BRITAIN'S GREATEST BRIDGES

Joseph Rogers

AMBERLEY

Acknowledgements

The author and publisher would like to thank the following people and organisations for permission to use copyright material in this book: Alessio Avezzano, Ellen Cross, Alan Frew, Pauline Gower, Steve Hoar, James Jones, Christine MacIntyre of Ardrishaig (Argyll), Alan Reynolds, Darren Rogers, Karen Rogers, Crosby Stone, Luke Walker, Mark Watson, Jonathan McDowell/Timothy Soar and Devon County Council.

In addition, the author and publisher would like to thank the British Library for their superb collection of hard-to-find books.

The image on the front cover is courtesy of Harshil Shah, www.flickr.com/photos harshilshah.

Every attempt has been made to seek the permission for copyright material used in this book. However, if we have inadvertently used copyright material without permission/acknowledgement, we apologise and we will make the necessary correction at the first opportunity.

First published 2019

Amberley Publishing
The Hill, Stroud
Gloucestershire, GL5 4EP

www.amberleybooks.com

Copyright © Joseph Rogers, 2019

The right of Joseph Rogers to be identified as the Author
of this work has been asserted in accordance with the
Copyrights, Designs and Patents Act 1988.

British Library Cataloguing in Publication Data.
A catalogue record for this book is available from the British Library.

ISBN 978 1 4456 8441 3 (print)
ISBN 978 1 4456 8442 0 (ebook)

Origination by Amberley Publishing.
Printed in Great Britain.

Contents

Introduction

When exploring the breadth and depth of Britain's varied landscape, a number of features make themselves known more frequently than others. For the urban adventurer, these features may come in the form of large-scale transport hubs, vast developments of housing or a significant industrial presence. Rural villages too have their own common features, such as vast acres of farmland, a central community hub, or outlying country estate. There are, however, some landmarks that present themselves in almost every settlement the nation has to offer, primarily churches, pubs and, rather regrettably, war memorials, but for a large percentage of locations, bridges are also prevalent.

This, of course, stems from our ancient British ancestors pursuing the logical steps in establishing a home in well-protected or profitable areas. Steep valleys provide shelter from the natural elements and force any encroaching enemies into restricted territory. Summits (likely above a valley) similarly are fantastic vantage points over weather and warfare, and allowed for simple forms of fortification. Rivers hold the necessary resources for water and therefore farming and sanitation, as well as act as a natural wall of defence.

Over time, overcoming these initially useful anomalies in the terrain became necessary, and as we progressed from early man to civilised human, the use of the surrounding landscape advanced from simply felling a tree across an impassable gap to learning basic mathematical principles in constructing crude wooden or stone structures that would last more than just a handful of crossings. Eventually, humanity's developing sense of exploration, conflict and trade meant that joining with settlements across a valley or beyond a river became more frequent, and with this increasing frequency came more technologically advanced bridging solutions.

In addition, bridges have stemmed from the need to carry other man-made creations. Today bridges not only carry people and livestock, but also motor vehicles, railway locomotives, utilities and canal routes. Bridges, along with religious buildings, have become icons of design, wealth and prowess, paying homage not to God, but to the great engineers that construct them, the royals and governors that authorise them and the people that ultimately use them. With such an appreciation for its vast and detailed history, today's United Kingdom exemplifies almost every age of British engineering and the desire to bridge gaps in the nation's landscape.

The following examples demonstrate key points in that history. Bridges that changed the way Britain approached such a technical feat or are noted for their development, relevance, aesthetics or lasting legacy. There are, however, innumerable structures that warrant in-depth analysis, poetic description or public recognition, most of which can only be found, appreciated or understood by seeing them first-hand.

Chapter 1

Tarr Steps, Exmoor

Prior to the technological advances in metallurgy, masonry and engineering, bridging consisted of utilising natural objects of considerable size to span a gap (a body of water more often than not). Early man may have used large tree trunks felled naturally, or taken them down purposefully. Equally, an accumulation of material such as smaller branches, stones and logs may have been used to dam a river or simply fill an area to a desired height.

In mainland Europe, some ancient bridges remain intact and provide examples of how some civilisations first tackled the issue of bridging. In Greece, the Arkadiko Bridge mainly features a collection of boulders moved into a ravine and was constructed predominantly in line with military movements. Further afield, in India, living bridges were guided across deeper ravines in the form of tree roots, which over time connected to opposing walls, and were then shaped and strengthened using a variety of techniques.

Notwithstanding numerous efforts to keep it in place, the bridge formed by Tarr Steps is surprisingly flat and stable. (Author)

'Tarr' may originate from the Celtic word Tochar, meaning causeway. This suggests that the bridge is indeed much older than most believe. (Author's Collection)

The origins of bridge building in Britain are hard to pinpoint. With many early structures long since washed away, dismantled or replaced, very few examples exist from ancient Britain. The oldest come in the form of clapper bridges, built across rugged terrain in parts of rural England and Wales using large, often uniform stones that are then positioned to form simple arches or spans. It is widely accepted that the majority of Britain's clapper bridges were made during medieval times, with the term itself coming from the Latin *claperius* meaning 'pile of stones', but such simplistic designs can be mistaken for products of a prehistoric nature.

Tarr Steps, situated in Exmoor National Park, is the largest clapper bridge in Britain, and though its age is almost impossible to verify, it is likely one of the oldest bridges in the country. With so much uncertainty about its origins, it is not surprising to hear that at one time locals considered it the devil's work, with those attempting to cross it threatened with death.

In reality, its use was likely associated with farming and the movement of livestock, people and packhorses across the moors. No fewer than seventeen slabs span the River Barle near Hawkridge, with a number of stones having been placed to deflect the flow of water and debris between crude arches during times of flooding. These stones were sourced locally from the Pickwell Down basement bed, a geological formation running from the coastline to the west through to nearby Dulverton, and vary between approximately 2 and 3 metres long.

Over time, substantial floods managed to dislodge the stones, carrying them up to 50 metres downstream. Following major damage in both 1952 and 2012, cabling has been put in place to assist the cutwaters in disposing debris and each stone has been recorded in a way that would aid engineers in rebuilding the structure, should such flooding happen in the future. Despite its age and simple construction, it still exists as a fully functioning bridge for pedestrians and adds to a number of mystical features within Somerset for which the origins are shrouded by myth.

Left and below: *Given its isolated location in a narrow valley of Exmoor, the scene around Tarr Steps has changed little over time. (Author and Author's Collection)*

Chapter 2

Clattern Bridge, Kingston upon Thames

Among one of the more unusual roles for Britain's bridges to take was that of a goal in a not-so-typical version of England's 'beautiful game'. Though the origins of Kingston upon Thames' Shrove Tuesday football match are somewhat unclear, its development into an annual fixture in the town added a great deal to the history behind one of the country's oldest bridges.

It was in the twelfth century, during the Norman period, that the current bridge was built over the Hogsmill River, replacing an earlier Saxon example known as the Clatrung Bridge. The name reportedly stemmed from the noise that resulted from horses crossing over it, perhaps more obvious in its later recording as the 'Clattering Bridge'. Three arches of flint comprised the structure, which formed a central roadway at the heart of the town and remained as built for around 500 years, before being modernised for heavier traffic in 1758.

Along with much of Kingston around the ancient market place, the bridge hints at the town's medieval heyday, with its success initially resulting from farming, brewing and tanning. Its proximity to the Thames, which was crossed by the more prominent

The scale of Kingston upon Thames' Shrove Tuesday football match is clear in this illustration. (Author's Collection)

Above left: *The original part of the bridge on the downstream side is now sandwiched between Kingston's varying architectural styles. (Author)*

Above right: *Upstream, a more recent facade presents the public along with a roadway much wider than its original 8 feet. (Author)*

Kingston Bridge, enabled the town to thrive thereafter and is now the reason for its importance as a key town on the outskirts of Greater London.

Shrove Tuesday's traditional football game, played for many years prior to the turn of the twentieth century, saw townsfolk aim to score in one of two goals – the Kingston Bridge and Clattern Bridge, supposedly with the aim of settling a dispute between Coenwulf of Mercia and his rival Keynard. Alternative origins for the game stem from a defeat of the Danish, in which the head of the Danish captain was kicked by the townsfolk of Kingston in celebratory fashion. Whatever the source, the tradition carried through the town in such riotous fashion that intervention was required by way of the Riot Act in 1798, before being ousted from the streets entirely by authorities in 1867 due to its disruption to trading in the town. Clattern Bridge's guise as a goal in this peculiar version of football may also have stemmed from its position on the outskirts of the town centre and those in support of the Danes may have sought refuge by fleeing here with the remains of their captain.

Outside of its place among exaggerated tradition, the bridge is now recognised as one of the oldest bridges in England, and with the aid of repairs and alterations in 1852, it still carries traffic through Kingston upon Thames today.

Chapter 3

Monnow Bridge, Monmouth

With bridges offering better access between settlements, across boundaries and obstructions, they too provided an advantage for those using them – an advantage that owners and authorities at times did not want to have available to enemies. The militarisation or fortification of bridges therefore prevented undesirables access into towns, estates or even entire countries, and allowed for protection against sabotage or other advances from opposing armies or rebellions.

Monmouth, a Welsh town on the border with England, seems to have lacked a permanent bridge across the River Monnow prior to the twelfth century, but as time progressed through the medieval period, the growing number of threats warranted not only replacing an existing wooden structure, but also fortifying it. Such drastic measures were likely prompted by the Battle of Monmouth in 1233, which may well have destroyed the wooden predecessor, thus necessitating a rebuild in stone. By around 1300, historians suggest that the bridge was complete with a gatehouse at its centre, though in a form differing from today.

Historians also debate whether the gatehouse was anything more than a glorified tollbooth. Those in favour of this conclusion point out that the Monnow could be crossed with ease on foot, though others point out that the expense and effort put into the structure far outstretches that typical for collecting tolls. Over the subsequent centuries

From this angle, the peculiar design of the gatehouse becomes apparent. (Author)

the gatehouse did hold a degree of military importance, from acting simply as a store for gunpowder and munitions to seeing through the events of the English Civil War of 1645, when it was seized by Royalists. Following this, the structure saw use as a gatekeeper's residence and a lock-up, as well as continuing to take tolls from those crossing it.

By the 1800s, as with a number of similar bridges in Europe, the gatehouse became something of an inconvenience for the increased levels of traffic. An extra arch was knocked through for pedestrians, but unlike other bridges, whose central buildings were under threat of demolition, the Monnow Bridge's gatehouse was never knowingly marked for removal. As a result, examples similar to Monmouth's fortified bridge became increasingly rare and eventually the feature of a central gatehouse was all but extinct in the UK.

Amazingly, road traffic continued to use the bridge well into the twentieth century, and despite a large number of accidents and fears of collapse following disregard for weight limits and unseen erosion to the piers, it was only in 2004 that a new bridge was opened to allow the Monnow Bridge to carry only pedestrians. Now unique in Britain, the bridge has seen interest from historians, artists, tourists and relevant authorities, and is now Grade I listed and a Scheduled Ancient Monument. The bridge is featured in a number of notable paintings, some of which lie in Tate Britain, and also acts a focal point for those travelling to Monmouth from the south.

From the grounds of Overmonnow's church, the bridge has changed little in over a hundred years. (Author and Author's Collection)

Chapter 4

Town Bridge, Bradford-on-Avon

The River Avon's history becomes more interesting as the river flows through Wiltshire and into North Somerset, particularly at Bradford-on-Avon, a condensed town once thriving at the centre of a prolific textile industry.

In order to better connect the town's people, a basic packhorse bridge was likely built sometime around the thirteenth century in place of an earlier bridge, near to a 'broad ford' from which the town got its name. It is known that this bridge was in some state of disrepair in 1400, prompting the Pope to ask those in faith to give alms to fund its repair, and though information is scarce following this, the likelihood is

Today's backdrop for Town Bridge is a series of redeveloped industrial buildings, now seeing use as quaint cafés, trinket stores and offices. (Author)

that the structure stayed largely unchanged until the seventeenth century. In 1540, the bridge is recorded as having nine arches, matching the number present today, though it is believed that only the two on the eastern side originate from the original thirteenth-century structure.

Bradford's involvement with textiles reached its height in the 1600s, using the force of the river to power a number of woollen mills. As traffic increased, larger loads were directed through the ford, while the rest continued to use the bridge. Widening the roadway to 25 feet 6 inches took place during this time to accommodate this increase and a small chapel was built over one of the altered cutwaters. All of the downstream arches on the western side were amended or replaced, with little of the older work visible. The quaint chapel was later used as a town lock-up, with drunkards said to be 'under the fish and over the river', in reference to the Bradford Gudgeon, a decorative, copper-gilt weather vane placed atop the building. Various records refer to the structure as the Blind House, possibly observing its omission of windows, and like many small buildings in similar positions, it may have acted as a toll house.

Such was the bridge's prominence in the centre of the town that its use refused to dwindle following the migration of the woollen industry to Yorkshire and elsewhere during the Industrial Revolution. Revisions were made to the cutwaters on the eastern side (some may not even have existed prior to the 1880s) and modern additions such as railings, street lamps and bollards were implemented. Now acting as the town's main thoroughfare northwards to Bath and south to Trowbridge and Westbury, it remains in its amended seventeenth-century format as it carries the A363. The redevelopment of many mills to residential and commercial purposes, married with the town's stance as a picturesque stopover for tourists, ensures the bridge's continued use as a focal point for those exploring the southernmost reaches of the Avon's course.

Chapter 5

Stirling Old Bridge, Stirling

Pinpointing the moment that first saw the River Forth crossed at Stirling has been notoriously difficult for historians. A number of dubious accounts, from as far back as the first century, make reference to the boundary at the Scottish settlement, but do not go as far as definitively confirming a structure spanning the river. It is widely believed, however, that a crossing, in one of form or another, existed during the early medieval period. Archaeological findings suggest this could have been as minimal as a ford, while others suggest modest wooden or stone structures.

Conflict, more often than not, had highlighted Stirling's strategic importance and hinted at bridging the Forth near to the town. A sixteenth-century account of the invasion by the Northumbrians and the Cumbrians in AD 855 reportedly resulted in celebration by erecting a crucifix on a stone bridge. Moreover, William the Conqueror

This engraving by C. Roberts depicts much drama at the Battle of Stirling Bridge as the structure falls beneath soldiers and cavalry. Some Scottish historians claim the collapse was on the order of Wallace, while others point the finger at Surrey. (Author's Collection)

The bridge sits on the north side of the city, in the direction of the National Wallace Monument. (Author)

is said to have crossed the river in 1072. The first legitimate reference of a bridge at Stirling comes during the 1290s, in the run up to and during the Battle of Stirling Bridge.

While the battle is remembered historically as the first fight for Scottish independence, and the resulting victory for William Wallace and Andrew Moray, its ability to confirm the existence of a bridge is also important. During the fight, the bridge (likely made of wood) was destroyed, possibly on the orders of the Earl of Surrey after it became clear that his English army had been defeated. The collapse of the bridge, laden with soldiers, is depicted in the late 1800s in dramatic fashion, likely as the result of the fictitious tales by the poet Blind Harry, and in later representations, such as the 1995 film *Braveheart*, the significance of the bridge is omitted entirely.

Today, only four stone piers lie below the surface where the ancient bridge might have stood. Downstream, Stirling Old Bridge now stands gracefully over the river, but is still of significant age. Completed in 1415 under the instruction of Robert III, this new bridge offered more versatility in providing a means for tolling and taxing as trade increased, as well as being wide enough to accommodate a guardhouse following repairs. In 1749, again as a result of conflict, part of the south side was rebuilt entirely, having been demolished to prevent Prince's Charles' army from advancing south, and the guardhouse too was eventually lost. This Stirling Bridge remained relevant right up until the Industrial Revolution, when it was succeeded by a more modern bridge that could cope with more traffic.

An image of William Wallace from the 1873 title British Battles on Land and Sea. (Author's Collection)

Chapter 6

Bridge House, Ambleside

The need to cross Stock Beck, a small river passing through Ambleside, Cumbria, was almost certainly a personal one, instigated by the local Braithwaite family over 300 years ago. Not only was there a desire to better access land on either side of the ghyll, but also to store a quantity of apples that resulted from the surrounding orchard. As a result, a rather unusual structure was constructed in the form of Bridge House, which, at a glance, seemed to be a small cottage-like building suspended over the water below.

This scene looks somewhat fantastical, with the surrounding trees and peaceful folk fishing. (Author's Collection)

It is widely accepted that the bridge was built first, with the house coming later. Besides the relatively basic need to cross the river and store apples, other more complex motives may have warranted the house's appearance. It is possible that Bridge House was a folly or summerhouse, built merely to compliment the grounds of Ambleside Hall, or that it was a peculiar way of avoiding land tax. As a result, and with a myriad of uses in the subsequent centuries, the precise origins of the house are difficult to pinpoint.

Its unusual design, however, has ensured interest among visitors to Ambleside, which is now popular with tourists visiting the Lake District National Park. Aesthetically, the structure bares much resemblance to its surroundings, with the harsh irregular rubble stones jutting out from one side and a shallow slate roof all sitting atop a gentle arch made from similarly coloured stone. Timber and ironwork furnish the interior, which over the centuries has been used to accommodate craftsmen and weavers, a tearoom and at one stage housed a family of eight.

Such was the house's place at the centre of Ambleside, and as one of its oldest and more notable buildings, that the locals decided to ensure its long-term survival during the 1920s. By this time, some stonework had deteriorated and repairs were much needed, so after considerable efforts to fundraise, these repairs were carried out and its future secured by the National Trust. In addition to the affectionate locals, many visitors had also gained an appreciation for the landmark, with a number of notable paintings depicting the beauty of its place among the nearby lakes and mountains.

With over eighteen million visitors to the Lake District per year according to Cumbria Tourism, Bridge House sits among Windermere and Scafell Pike as one of the area's major tourist draws, putting Ambleside firmly in the limelight. Despite its humble and confused beginnings as a personal endeavour for the Braithwaites, it is now a significant part of the town's very existence as a tourist destination.

Chapter 7

Packhorse Bridge, Carrbridge

The concept of bridging a waterway, namely a river, signifies a need or want to overcome one of nature's physical barriers. In some areas, and under certain conditions, it would seem that nature wishes to undo humanity's efforts. One such area is the Scottish Highlands, where vicious torrents and wide firths provide considerable challenges for those wishing to conquer them.

While the River Dulnain, a tributary of the River Spey, seems a relatively minor feature to cross in places, its course running through the Monadhliath Mountains does pose occasional problems in the way that it flows. At Carrbridge, a gateway to more adventurous landscapes in the form of Cairngorms National Park, the Dulnain narrows and passes through a bottleneck prone to harsh currents during torrid conditions. This mood was often described as the river being 'in spate'.

This did not stop local communities building bridges however, and in 1717 a packhorse bridge was built across the Dulnain at Carrbridge to give funeral processions access to the nearby Duthil Church. John Niccelsone of Ballindaloch, a local mason, was financed by Brigadier-General Alexander Grant and created the bridge roughly midway between two right-angled meanders. This provided much better access to the area for tradesmen, travellers and local people, as few alternative crossings existed.

Above: *Robert Burns (1759–96).* (*Author's Collection*)

Left: *Storms, floods and torrents were not uncommon for Caledonia. Robert Burns described such events in his 1781 poem* 'Winter: A Dirge'. (*Author's Collection*)

The river tumbles over rocks and stones before passing under the now dainty remains of the bridge. (Mark Watson)

The mood of the Dulnain shifted to violence during 1829, when a series of floods along some of the region's larger rivers created the 'Muckle Spate'. On 2 August, tremendous thunderstorms materialised over the Cairngorms, causing mass downpours and considerable run-off for smaller tributaries and streams to cope with. The River Dee, running east, reportedly rose to 15 feet, with other waterways similarly getting saturated beyond their normal levels. By the early hours of the following day, some bridges had been washed away completely, with the bridge at Carrbridge surviving, but with considerable damage to much of its structure.

Its resulting form, fragile and broken, somehow matched its status as the oldest bridge in the Highlands, and with the settlement in time becoming a focal point for curious hikers and mountaineers, the bridge too gained notoriety as a local landmark. The exposed and feeble stonework, together with a slender, almost circular arch and idyllic setting, see the unused bridge admired by many despite the loss of its use as the 'coffin bridge'.

Chapter 8

Wooden Bridge, Cambridge

As the need to span more gaps and obstacles increased over time, the technological aspect of bridge-building became evermore complex. With their famously straight roads, the Romans used mathematics to pave their way across Europe, similarly applying it to the structures needed to cross rivers and valleys, and as the understanding of physics came to light, load bearings, weight distribution and the science behind building materials meant that bridges developed into products of substantial labour and research.

As one of the most renowned and successful universities in the world, it is no surprise that Cambridge has been at the heart of ground-breaking leaps in the understanding of bridge-building fundamentals. Sir Isaac Newton, one of the more famous alumni of Trinity College, laid the foundations for a number of mathematical principles, and while elements of his study bear little relevance to engineering bridges, such as his theoretical calculation of the speed of sound, others such as understanding his eponymous Newtonian (and more interestingly non-Newtonian) fluids have helped understand how stresses and forces alter the physical properties of some materials. More importantly, Newton's progress with Gottfried Leibniz in mathematics, furthering the work of ancient academics, resulted in principles crucial in making some of the more complex calculations needed in bridge construction.

In 1749 an architect by the name of William Etheridge designed a bridge that candidly exposed its underlying mathematical principles. Using a series of straight wooden beams to create tangents forming an arc, Etheridge demonstrated the elegant skeleton upon

Each pair of central radials is separated by an angle equal to 11.25 degrees, or 1/32 of a revolution. In addition, the tangents form an arc with radius of 32 feet. (Author's Collection)

The bridge now links a more contemporary part of the college with its older buildings. (Author)

which most stone bridges of the time were laid. The resulting simple structure, placed initially at Trinity College before being moved to cross the River Cam at Queens' College, demonstrated tangent and radial trussing, a concept making use of the physical forces involved in such structural arrangements. Etheridge applied the same methods to his Old Walton Bridge at Walton-on-Thames but on a much larger scale, setting the scene for Canaletto's painting *Old Walton Bridge* of 1754.

Today the 'Mathematical Bridge', as it is more popularly known, acts as an advertisement for Cambridge's academic prowess. Though all entirely false, numerous myths related to Newton's involvement with the design, influence from China and that the bridge requires no pins or bolts to stay erect create an air of mystery around the fact that mathematics can manifest itself so clearly in a physical object.

Old Walton Bridge demonstrated the same mathematical principles, but on a much larger scale. (Author's Collection)

Chapter 9

Pulteney Bridge, Bath

The influence of Italian architecture became clear throughout the late 1700s as Bath continued to grow. A desire from the wealthy Pulteney family to connect the city with their nearby rural estate at Bathwick required bridging the Avon, and it was inspiration from Florence that saw architect Robert Adam implement rows of shops along both sides of the resulting structure. This novel feature was in fact once commonplace for busy, innercity spans, and while Adam had no doubt taken elements from the Ponte Vecchio in central Florence, other notable examples included London Bridge, which at the time was having its shops demolished to ease traffic, Bristol Bridge, which was undergoing similar revisions, and the Irgandı Bridge in Busra, Turkey. Further to this, Andrea Palladio's style was again implemented for a bridge in Bath, but this time with real purpose in mind and not merely as an exercise in exuberance, such as Ralph Allen's Palladian bridge at Prior Park.

The construction of Pulteney Bridge began in 1770, and though the bridge and commercial properties atop were built by separate masons, the final structure was nevertheless one project. Its completion four years later marked the beginning of numerous alterations by shop owners and local authorities, and by the start of the twentieth century, like London Bridge, the commercial use of the bridge had been abused, detracting from its architectural merit. Within thirty years, parts of the structure had begun to fail, warranting repairs to the central stonework and a rejected proposition by Thomas Telford to replace the bridge entirely. Post-war restoration ensured the survival of the bridge's Palladian spirit on the south side, now an iconic image of present-day Bath's UNESCO World Heritage site status. The north side, having had its shops rebuilt in 1802, lies conveniently behind a contemporary supermarket, hidden to the majority of visitors, but still demonstrates the manner with which trade imposed upon the design. Plainly rendered extensions hang unevenly over the river and do little to blend in with

At one time, crowding bridges with houses and shops was standard practice in cities. Many, like this example in Newcastle, have since been replaced. (Author's Collection)

the grand Georgian style of the rest of the city. The prominence of Pulteney Bridge's favourable face looming over the Avon and niche, specialist dealers lining the street at its centre ensure its fulfilment of purpose, along with allowing Bath to grow eastwards and expand its much applauded, consistent style.

Above: *The other side of the bridge has alterations and additions adorning most of the rear façade. (Author)*

Right: *The Italian influence is seen best from a low position by the river. Note the weir's straight edge, which was altered to a sharp arc by 1972. (Alan Reynolds' Collection)*

This view from the roof of the Empire Hotel represents an example of uniformity among Bath's buildings. (Alan Reynolds' Collection)

Chapter 10

Countess Wear Bridge, Exeter

A certain amount of greed was arguably to blame for the development of the River Exe near Topsham in the thirteenth century. In owning the estates nearby, the Countess of Devon (or possibly her relatives) sought to limit river traffic upstream to Exeter, wanting to make profits from having merchants and traders offload their cargo at Topsham instead. Though dates vary, it was around 1284 that an obstruction across the river was placed in the form of a weir, giving what is now a suburb of Exeter its current name.

When the time came (approximately 300 years later) to undo the work imposed by the countess, it proved to be somewhat of a challenge, with the creation of the Exeter Ship Canal offering an easier solution by simply bypassing the river. With navigation of

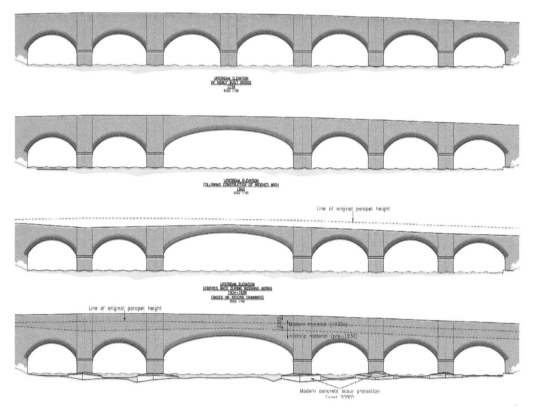

In improving the traffic situation on Exeter's periphery, those involved sought to chart the history of the Countess Wear Bridge, so that any alteration would be done with consideration. (Devon County Council/Luke Walker)

the Exe now effectively redundant, little stopped the construction of a multi-span bridge in the weir's place, and by 1774 an attractive road bridge was opened. Construction came through local man Thomas Parker who, with aid from the Exeter Turnpike Trust, created a bridge of limestone rubble and red sandstone. The parapets featured triangular recesses extending up from the cutwaters to form refuges, and though each of the seven spans were seemingly of uniform width, they in fact varied between 23 and 26 feet.

In 1842, the central arch was widened by means of knocking through one of the piers and joining it with the adjacent arch to the east, giving the structure the basis for its present shape. A hundred years later, and with a vast increase in traffic, significant works were once again performed, widening and raising the roadway and parapets. This enabled the bridge to take on Exeter's traffic in the form of a bypass, diverting vehicles away from the city centre, across the Exe and onwards further south.

The introduction of the M5 motorway in 1977 enabled traffic to bypass Exeter entirely, however, and so with the expansion of the city's suburbs and industrial estates, the bridge's function changed once again, this time aiding the flow of traffic around Exeter's outskirts. For many years, the area was considered a bottleneck, with solutions seemingly bound to removing the combined footpath and cycleway to increase the number of lanes on Bridge Road.

A proposal for repositioning the footpath and cycleway onto a new bridge adjacent to the northern parapet was granted initially in 2009, and by 2014 significant work in creating a modern, unobtrusive structure had commenced. In August of 2017, the finished article, designed and overseen by Devon County Council Engineer Luke Walker, was finally opened to the public along with a new road layout consisting of four lanes for traffic. Though intimately linked with the Countess Wear Bridge, the new structure forms a footbridge in its own right and by being considerate in its design, actively compliments the history of its Grade II listed partner in solving a contemporary issue for Exeter's flow of traffic.

Above: *To onlookers from the riverside, the newly introduced footbridge runs seamlessly against the old – a prime example of modern infrastructure that's sympathetic to the past. (Luke Walker)*

Right: *The addition of the footbridge resulted in a series of road closures, but it was essential in keeping all forms of traffic flowing around Exeter's periphery. (Luke Walker)*

Chapter 11

Iron Bridge, Shropshire

Iron is one of the most important elements on the periodic table, particularly when it comes to the development of humanity. Its existence on earth, as the most abundant element, has origins quite literally among the stars, where it is distributed at the climax of their life cycle across space and ultimately into the cores of rocky planets such as our own. Biologically, iron is also important, playing its part in the production of haemoglobin, which is essential for ensuring oxygen is carried throughout the body via the bloodstream.

Together, as both a tangible resource beneath the ground and as a means for human life to exist, iron takes form as a tool to be extracted, forged, shaped and utilised. Ancient craftsmen, whether situated in Asia, Europe or the Middle East, used iron found from meteorites to create jewellery and weapons, and as an understanding of obtaining iron from ore developed, civilisations began to work the metal for a variety of uses.

By the 1700s, ironworks were developing hand-in-hand with the coal industry as both resources were used in mass-producing the metal in the form of cheap cast iron. Though the Chinese had been using such methods for over 2,000 years, it was Europe's ambition to produce the material faster and cheaper that sparked the ensuing Industrial Revolution. At Coalbrookdale in Shropshire, the combination of successful coal mining and iron smelting saw the metal being used in a staggering variety of settings, most notably in replacing expensive brass in early steam engines.

Abraham Darby's blast furnace at Coalbrookdale was responsible for much of the pioneering work around mass-produced iron, and by the time the need arose for the

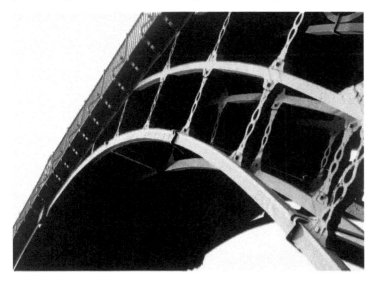

Upon closer inspection, the ironwork, at over 200 years old, is pitted and worn in places. Extensive restoration work began in 2017. (Steve Hoar)

Above: *The Iron Bridge showcases the decorative and elegant qualities of the metal. (Steve Hoar)*

Right: *Iron bridges became a common sight following the first one at Coalbrookdale. The Burnet Patch Bridge in Exeter is small in comparison, bridging a small passageway. (Author)*

nearby Severn Gorge to be bridged, some sixty years after his death, his iron was understood enough to be considered a suitable building material. With his grandson, Abraham Darby III, involved in the project, construction began in 1777 on what would be the first cast-iron bridge in the world.

Over 380 tons of iron were used in a design similar to that of wooden bridges, using a network of skeletal elements and components to form a graceful arch across the river below. When completed in 1779, the Iron Bridge marked the start of iron's use in larger construction projects, with Thomas Telford in particular making use of the material in his ambitious and impressive future feats. Now a Scheduled Ancient Monument, the bridge has undergone a number of restorations to ensure its iconic status lives on, and though its ability to carry traffic across the renamed Ironbridge Gorge has dwindled, it is still arguably the most important bridge in the country.

Chapter 12

Clachan Bridge, Isle of Seil

A key feature of the 790 offshore islands that surround the Scottish mainland is the manner in which they are separated. For some such as Arran and Colonsay, aircraft and ferry services alone facilitate the connection to the rest of Britain, but for those within a stone's throw, bridging short stretches of sea creates a permanent link.

This link was made for the Island of Seil in 1792 when Robert Mylne, known largely for his work on the first Blackfriars Bridge, built the Clachan Bridge 10 miles south-west of Oban. To accommodate ships, a single arch was favoured over an original proposal for a double arch, with a narrow roadway travelling over it. Rubble was used in the construction, giving it an attractive tinge in addition to its high position over the Clachan Sound. Initially this link was well received and enabled islanders from the main settlement of Balvicar to better trade and travel.

However, a dispute in 2011 over government funding resulted directly from the existence of the bridge. The Special Islands Needs Allowance, worth up to £400,000 for the island, was withdrawn following the decision to only award the grant to islands without a physical connection by road. This led to the Scottish press declaring that the government had effectively dismissed Seil as an island, making it part of the mainland. The reality is that it was only with regard to the grant that Seil was no longer considered an island. Geographically being separated by a body water, the Isle of Seil has its status in the name, with the locals still being considered islanders by the rest of the country.

Robert Mylne's first major project was the William Pitt Bridge, more commonly known as the first Blackfriars Bridge. (Author's Collection)

Further discussion stems from the bridge's unofficial titles of the Atlantic Bridge or Bridge Over the Atlantic. While the structure only spans the Clachan Sound, it has been argued that its titles are entirely justifiable as the Sound could be considered part of the Atlantic Ocean. Today the bridge carries the B844 road onto the island and is noted by locals and tourists for its display of fairy foxglove (*Erinus alpinus*) on its southern face, which flourishes in purple during the early summer. With a significant number of homes on the island now let for holidaymakers, the Clachan Bridge continues to allow Seil to regularly interact with the mainland.

Above: *A vibrant purple tinge, in the form of fairy foxglove, lines the southern parapet during the summer. (Christine MacIntyre of Ardrishaig, Argyll)*

Right: *At other times in the year, the same scene looks decidedly bleak. (Ellen Cross)*

Chapter 13

Skerne Bridge, Darlington

Upon the realisation that railways were to be the next stage in the development of transport, numerous roles and industries arose to satisfy this latest field of technology. With the railway revolution about to commence across Britain, it was a gentleman by the name of Ignatius Bonomi who would arguably be the first to take the title of Railway Architect.

Previously, Bonomi had given much of his architectural expertise to County Durham in the form of restoration work, first started by his father Joseph, and notable buildings such as Egglestone Hall. As construction began on the Stockton & Darlington Railway in 1821, and with George Stephenson keen to see his steam locomotive surpass the power of horse-drawn carts, it was clear that substantial engineering feats were required at certain parts of the route. One such location was in crossing the River Skerne, and as a bridge surveyor, Bonomi found himself offering advice to Stephenson.

Above left: *George Stephenson (1781–1848). (Author's Collection)*

Above right: *Today's cycle path threads through one of the flanking arches. The wall to the left remains from when the bridge was much wider. (Author)*

Bonomi expressed a desire to use coursed sandstone over iron, with a large central arch flanked by two smaller arches. Given a rise in the price of iron, this was accepted, and so in 1824 the ravine's sides were initially filled with rubble before construction began on one of the world's first railway bridges. By September 1825 services began, but suffered teething problems through the reliability of early steam locomotives. Once in its stride, the railways fed a ravenous increase in industry, and as a result the bridge was widened to satisfy more railway lines. Darlington itself became almost reliant on the new mode of transport, as demonstrated by the way in which the railways surrounded the town, threading outwards to the rest of the country.

The bridge was repaired not soon after it had opened. Cracks became evident near the embankments, which were slowly subsiding, and work was carried out to rectify this. Eventually, the northern face was dismantled, reverting the structure back to its original size with only the supports of the widened section remaining. Its current appearance reflects its likely status as the oldest railway bridge in continual use. The parapet is damaged, with other parts noticeably uneven, but nevertheless the Skerne Railway Bridge stands strong against the flow of Class 142 Pacers heading to Bishops Auckland and Hitatchi rolling stock emerging from the assembly plant at Newton Aycliffe – a far cry from Stephenson's *Locomotion No. 1.*

The structure received Scheduled Monument status in 1970 and was later featured on the reverse of the £5 note along with George Stephenson, whose recognition as the father of the railways somewhat eclipsed the efforts made by his advisor Bonomi.

After years of neglect, the bridge is now celebrated as a local landmark. (Author)

Chapter 14

Menai Suspension Bridge, Anglesey

The Isle of Anglesey, for much of its history, had been dependent on the sale of cattle to market towns on the Welsh mainland and beyond. As a result, farmers often had to organise the incredible feat of droving livestock through the Menai Strait, a body of water boasting fast-flowing currents and treacherous whirlpools. By 1790, despite the introduction of ferries, over 10,000 cattle per year were guided through the water, and with the inevitability of losing livestock to the tide, locals were keen to hear of a possible alternative.

It was the possibility of linking Anglesey with Ireland, however, that prompted a serious look into bridging the Strait, with Thomas Telford seeing the value in creating a route from London to Holyhead, so that ferries could link the English capital with Dublin. During the late 1700s, an increasing number of coaches were making their way across Britain to access the Emerald Isle and voiced concern about the safety

Arguably the best view of the Menai Bridge is from Church Island to the west. The graveyard on the island contains a memorial to those killed during its construction who weren't given official burials. (Author's Collection)

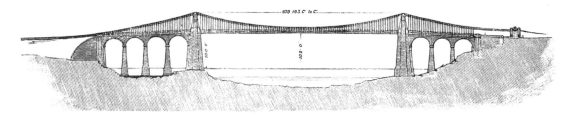

A diagram showing some basic dimensions of the bridge, as viewed from the south-west. (Author's Collection)

and reliability of the Menai Strait ferries. Naturally, a bridge would favour travellers from London and the locals in assuring a safer passage above the water and bring more opportunities for trade, travel and work. Objections were strong from the maritime community, who rightly pointed out that up to 4,000 ships sailed through the Strait each year, so any bridge would need to stand high above the water. In 1818, permission was finally granted for a suspension bridge of Telford's design to span the water at Ynys y Moch, a small island from which farmers historically began their droving across the Strait.

A huge amount of activity resulted from the construction of the bridge, which ranged from the flattening of Ynys y Moch itself, to installing railways, kilns,

Penmon limestone was not the first choice of material for the two bridge towers; original sketches show that the use of iron was originally planned. (Author)

35

workshops and offices on both sides of the shore. Up to 200 men found themselves at work and the population of Porthaethwy (since renamed as Menai Bridge) increased fourfold with a small number of them perishing as it was built. Stone was sourced from Penmon to create the arches and towers on each side and the iron chains weighed 25 tons each. By January 1826, the bridge was complete and stood proudly between the embankments either side, 100 feet above the water, as Britain's largest bridge.

While its size has now been surpassed many times over and the Menai Strait spanned once more by Robert Stephenson's Britannia Bridge, the importance of the Menai Suspension Bridge to Anglesey still remains. Gone are the days of wading cattle through to Bangor, with the tourism industry instead bringing people across the Strait. Looking from the west, the bridge has the immaculate backdrop of Snowdonia National Park, and was considered for UNESCO World Heritage status in 2005. Now, Anglesey is very much thought of as an integral part of North Wales and owes some of its popularity to its physical connection to the mainland.

The Bridgemasters House on the Welsh mainland holds the anchoring point for the two sets of steel chains, which replaced the wrought-iron ones in 1938. (Author)

Chapter 15

Galton Bridge, Smethwick

The historic, industrial heart of the West Midlands lies within an area known as the Black Country. As Britain began to exploit the land and machine the environment on an unprecedented scale throughout the nineteenth century, it was this part of England, north-west of Birmingham, that produced a significant amount of labour-driven tools and resources. At the same time, the surrounding infrastructure, growing with the introduction of a vast canal network and equally impressive railway lines, criss-crossed a landscape pitted and scarred with the upheaval of coal and lit by burning furnaces and foundries at night.

In this truly artificial setting, bridging a gap, valley or trough didn't necessarily mean spanning a natural feature, and when the need came to expand Smethwick, a town outside of the Black Country's borders, towards West Bromwich, it was the previously placed canal that posed an obstacle. Thomas Telford, now with almost thirty years of successful bridge-building experience, designed a simple iron structure to cross over 150 feet in a

Above left: *Telford's Galton Bridge sits impressively above the canal below. (Darren Rogers)*

Above right: *The roadway on top has become absorbed by the surrounding infrastructure of the West Midlands to become nothing more than a plain crossing for pedestrians. (Darren Rogers)*

single span, 68 feet above the canal. All 700 tons of the iron used had little distance to travel, being cast by the Horseley Works at Tipton, less than 5 miles away. Horseley Ironworks had already graced the local canal network with its popular type of roving bridge, but had also made its mark in the Smethwick area with the Engine Arm Aquaduct four years earlier.

Construction started on 30 June 1828, and after less than a year the bridge was complete. The routine aspect of bridge-building for Telford and iron casting for the Black Country meant that while Galton Bridge, named after Lunar Society member and arms manufacturer Samuel Galton, was reportedly the highest in the world. Over time, though, its practical function eclipsed its importance. The use of Black Country iron was celebrated, however, by Joseph Reeves in his 1836 book *The History & Topography of Westbromwich and Its Vicinity*. He points out that the building of such bridges, compared to the creation of Egypt's pyramids 'discovers an application for the smelting of iron, and casting in its moulds, adapted to those useful purposes, of which the ancients were comparatively ignorant and uninformed'.

Bridging mankind's previous endeavours became a necessity for the Smethwick area. Within the square mile surrounding Smethwick Galton Bridge railway station, built in 1995, no fewer than twenty road, foot, water and rail bridges wind their way through each other in almost every combination. Though the industrial might of the Black Country and adjacent towns such as Smethwick has declined over time, the need to move swiftly between settlements has continued to grow and with it the need to traverse older or alternative means of transport.

The platforms of Smethwick Galton Bridge railway station are themselves carried across an adjacent bridge. (Darren Rogers)

Chapter 16

Marlow Bridge

In 1849, the Széchenyi Chain Bridge opened after nine years of construction, linking the Hungarian capital of Buda with Pest on the other side of the River Danube. For Hungarians, it was a hugely significant feat that demonstrated ambition for the nation despite a failed revolution in the previous year. As the first permanent crossing in what would later be known as Budapest, it marked an increase in fortunes and development for the city.

Its origins were to be found in Britain, however, where William Tierney Clark was progressing as a notable civil engineer. In the late 1820s, he established a suspension bridge across the River Thames at Hammersmith, in what was an example of pleasing proportions. The stone towers were accompanied by a road deck made from granite and timber, wrought-iron chains and octagonal toll houses, and though the river's latest crossing was initially seen as a success for Clark, its practical use at a time of ever-increasing traffic was deemed insufficient less than fifty years later.

However, Clark's attractive design did continue further upstream with the building of a bridge at Marlow in 1832, where previous attempts to span the Thames had resulted in collapse or destruction. Again, square towers with an arch guiding the roadway through were implemented with chains suspending the deck above the water. Following the erection of his second suspension bridge, Clark was approached for an initiative by István Széchenyi, who saw the spanning of the Danube at Buda as an opportunity to progress his ambition to make the Hungarian capital a central hub for economics, culture and politics in Europe. His ties with Scottish engineer Alan Clark (to whom

Rowers enter the water with Clark's Hammersmith Bridge in the background. Spectators for the University Boat Race of 1870 prompted concern over its ability to bear weight, hence being replaced by Joseph Bazalgette's bridge shortly afterwards. (Author's Collection)

Above: *In applying an up-scaled version of his Hammersmith and Marlow bridges to Buda and Pest, Clark created an enduring masterpiece for the European capital. (Author)*

Left: *Marlow's quaint British landscape couldn't differ more from the Budapest of today, even with their bridges being similar. (James Jones)*

William had no relation) saw vast improvements in the city's infrastructure, and though the concept of a bridge had benefits in transportation, it was the grand symbolism that motivated Széchenyi's desire for one.

For a third time, William Tierney Clark chose his familiar layout for the project and had the bridge built in the UK before shipping it directly to Hungary. As expected, the structure drew much applause, cementing Széchenyi's place in Hungarian history and justifying Clark's design in a setting where form came above function. The bridge, later named after Széchenyi, soon became the centrepiece of a unified Budapest and was painstakingly restored after being demolished by German forces during the Second World War. Today, the bridge is an icon at the centre of one of Europe's most beautiful cities.

Similarly, its small-scale predecessor in Marlow is seen as a success. While its practical use has again been surpassed, its place as the concept of one of Europe's most important bridges makes Marlow Bridge, and its Hungarian counterpart, arguably two of Britain's greatest.

Chapter 17

High Level Bridge, Newcastle

Today, no fewer than seven bridges cross the River Tyne within the area south of Newcastle's thriving city centre. Among these are examples carrying the main line railway, major roads, foot passengers and utilities, with the arched Tyne Bridge and Gateshead Millennium Bridge fixed in most people's minds as the most iconic. Certainly aesthetically, both create a past-meets-present message relevant to Newcastle's status as a modern British city, but further upstream the North's grounded, practical achievements are apparent in another notable structure.

During the mid-1830s, the requirements at the point adjacent to the city's namesake were for road and rail to pass over the Tyne simultaneously, with an emphasis on the former and the latter mainly acting as a means to gain revenue. There was also an increasing desire to link Scotland with London and other major English cities via the growing railway network, and though a number of low crossings for road traffic had been in place since AD 122, a further span was seen to be necessary. Initially, this combined use was to see road and rail placed side-by-side, but after restrictions were realised by the chief designer of the project, Robert Stephenson, a decision to place the railway line above a roadway was made.

Construction began in 1847, with considerable effort being made to overcome the soft sand and changing tides as they impacted the laying of foundations. Only eighteen months later, the first train was seen gingerly making its way over using a temporary line, and after examinations were made in 1849 by the Inspecting Office for the Board of Trade using a train with fully laden wagons and four locomotives, the High Level Bridge was opened to passenger services. A lavish inauguration was conducted by Queen Victoria on 28 September, bringing much of the city to a halt. Shops and factories

The equally substantial Swing Bridge, opened in 1876, aided the flow of traffic along the Tyne, as depicted here. Its use, along with the High Level Bridge, saw river, rail and road add to the vast infrastructure of Newcastle and Gateshead. (Author's Collection)

were instructed to cease activity by the mayor and crowds were seen to gather many hours before the scheduled arrival of the royal train, despite inclement weather.

The primary focus of fanfare centred on the railway's expansion across the Tyne, and though the tolled roadway beneath had similarly been opened to the public, its use was somewhat underwhelming. The introduction of tramways in both Newcastle and Gateshead saw both decks adorned with rails for a short period from 1922, but for much of time the lower deck was the haunt of horse-drawn buses and their motorised successors.

Though its existence stemmed from the practical need to carry vastly improved transport infrastructure in and out of Newcastle, the High Level Bridge stands today as one of the city's most iconic engineering feats. Trains still run across the upper deck despite the massive increase in weight of most locomotives and the lower deck still allows for pedestrians and bus users to admire the Tyne from a commanding position.

Newcastle from St Mary's

Above: *This view, from St Mary's in the early 1920s, shows the prominence of the High Level Bridge in guiding road and rail past the castle from which the city gets its name. (Author's Collection)*

Left: *From the riverside, the stature of the High Level Bridge does its name justice, with the railway deck at 120 feet above the water. (Author)*

Chapter 18

Britannia Bridge, Anglesey

During the early nineteenth century, the desire to connect London with Dublin continued to grow beyond the needs of agriculture, and though Telford's suspension bridge allowed for the passage of horse-drawn coaches, the expansion of the railways presented yet another opportunity for both Anglesey and Britain's trading relationship with Ireland to expand.

In 1838, George Stephenson opined that using the existing single-carriageway Menai Suspension Bridge for a railway crossing was unsuitable given the weight required for heavy locomotives. Following that, his son, Robert, was appointed chief engineer on a project to build a second purpose-built bridge for the railway a mile west of the first crossing. In contrast to the elegant curves in the arches and chains of Telford's bridge, Stephenson opted for an angular box design that would see a central stone support flanked by two smaller ones and linked with box girders containing the tracks. Simultaneously, the box girder design was being implemented at Conwy Castle, where a combination of Telford and Stephenson's ingenuity had similarly facilitated better access along the North Wales coast.

Great care was taken to ensure that the tubular, iron structure of the girders could support the weight of locomotives and more. William Fairbairn, upon analysing the concept, insisted on the girders being over-engineered and regarded the need for additional support from suspension chains as 'highly improper'.

After opening in 1850, rail services ran through from London Euston to Holyhead, where growing port infrastructure allowed for a switch from train to ferry in less than two minutes. Greater access to Dublin for larger quantities of freight and passengers was gained, building on the success of the initial crossing twenty-four years earlier. As steam power waned and

From this angle, sharp lines dictate the shape of the Britannia Bridge, with cubic tubes spanning the block-like towers in the Menai Strait below. (Author's Collection)

This somewhat stylised depiction of the bridge shows onlookers viewing busy traffic along the Menai Strait. (Author's Collection)

the diesel locomotive age took hold over the subsequent century, the bridge continued to provide Anglesey with an active role in Britain's transport link with Ireland.

In 1970, however, that link was temporarily broken. Boys on a nature expedition inside the iron girders caused a huge fire when they lit a flame for illumination and ignited the tarred, hessian material where the iron structure met stone. As pointed out in a report by the Caernarvonshire Fire Brigade, with tar coating the tubes and a breeze along the Strait, the initially small fire quickly raged out of control.

By the morning of 24 May, the damage to the bridge had been done, and an entire rebuild commenced – aside from the stone columns, which had survived the inferno. A concrete deck for the railway replaced the charred iron tubes and was supported by arched steel in a style befitting the age the bridge was built in, not repaired in. In fact, when the bridge was reopened, complete with the addition of a second, higher road deck carrying the A55, it looked rather more pleasing to many than the original structure. Overall, the rebuild by Husband & Co. was deemed a success in securing the future of an important structure damaged by misfortune and improving it in line with the increase of road traffic in the mid-twentieth century. Once again, the demand for Anglesey had been met by ambitious engineering, which to this day provides access to north-west Wales.

Today's elegant arches beneath the railway deck show tasteful sympathy from the 1970s, seldom seen during a period known for its Brutalist movement. (Author)

Chapter 19

Scamells Bridge, Salisbury

Abraham Darby III's breakthrough of building bridges from iron eventually saw examples being produced en masse like any other commodity made from the metal. By the 1840s, Joseph Butler & Co., based at Stanningley near Leeds, were busy producing large structural items for the railways including parts of York railway station, one of the two Gauxholme Viaducts and a bridge for the Leeds & Selby Railway in 1848, all from iron. Three years later and a few hundred miles further south, the firm was placing another iron bridge across the River Otter in Devon, this time serving as a road bridge for traffic travelling between Ottery St Mary and Exeter.

By 1857, and after a number of understated, yet important iron products to their name, Joseph Butler & Co. were tasked with supplying castings for a bridge over Castle Street in Salisbury to allow the railway to run through the city and then out, heading toward Yeovil and Exeter. The city was keen to integrate the railway, with one unknown Salisbury local being quoted in the press as saying that 'Salisbury might become the Manchester of the South'. The bridge was much shorter than the one in Devon and featured four arched girders.

For reasons largely unknown, but likely involving some form of structural uncertainty or inadequacy, the small railway bridge was moved and replaced by a much less appealing example. At the time, such items were usually melted down, but Butler's bridge over Castle Street saw a second life as a road bridge. Its new location was to be less than 100 metres away, across the Salisbury Avon (not to be confused with its more renowned Bristol counterpart). Remarkably, despite the availability of innovative machines and construction apparatus, the bridge was moved without such aids and it

Butler's prior achievement at Ottery St Mary seems nondescript from the roadway, but rather grand from the riverbank below. (Author)

and its amendments were proudly positioned by hand by 1898, some forty years after being built. Concrete piers at either side, along with iron latticework, renewed the structure for road use and saw the bridge being named after its transporter, T. Scamell, and not after its creators in Leeds.

The 1975 *AA Touring Guide* referred to the A3393 running through Salisbury, and it was with this road that Castle Street's railway bridge initially found a new purpose. In time, larger roads succeeded in moving traffic in and around the city, leaving the bridge obsolete.

Today, the somewhat neglected Scamells Bridge (sometimes spelt with an additional 'm') sits as part of Nelson Road, crammed between terraced housing, the swirling traffic of the A36 and the railway that it once formed an intimate part of. A suitable plaque commemorates efforts in moving the bridge, but fails to mention its origins as part of an industry mass-producing elements of the transport network from iron. Only onlookers from the cycle path below can observe the plain arched lettering on the side, mentioning Butler and the Stanningley Ironworks. In contrast, revisions to Joseph Butler's bridge at Ottery St Mary in 1992 were widely celebrated, and though much of the structure beneath now consists of concrete and steel, the iron facade and structure's scheduling as an Ancient Monument remains as evidence for the firm's place in Britain's bridging history.

On-street parking for Nelson Road is the bridge's primary purpose now, with little indication from the road of either its age or previous life. (Author)

Chapter 20

Royal Albert Bridge, Saltash

Isambard Kingdom Brunel's journey south with his bold and innovative expansion of the railways wasn't an easy endeavour. Prior to his involvement in engineering the lines south-west of Exeter, much debate centred on the desired route towards the narrow county of Cornwall. Two railway companies offered their proposals, with the London & South Western Railway seeing an easier route north of the peaks of Dartmoor and the Cornwall Railway suggesting a route along the coast that would bring a rail service to more passengers at some of Devon's larger coastal towns.

The latter came to fruition, but only following the backing of the Great Western Railway and Brunel taking responsibility as engineer. Aside from the challenge of running the track against the sea, where time would dictate relentless coastal erosion, Brunel decided to trial his version of the doomed and costly Atmospheric Railway concept when negotiating the route down to Plymouth. After less than a year of

A rather dramatic view of the Royal Albert Bridge is seen here in A. Ansted's drawing of a sketch by J. L. L. W. Page. (Author's Collection)

operation, the result was impractical and unpopular; however, under steam power, the railway fared better, with the increase of Devon's tourism industry and holidaymakers flooding to the newly dubbed 'English Riviera'.

The next step required bridging the River Tamar at Plymouth and connecting the Cornish people to Britain's expanding transport network. The Royal Navy's presence at Devonport meant that restrictions were imposed as they exercised their responsibility to ensure the navigability of Britain's waters. After proposals to carry trains across the Hamoaze at Torpoint by ferry were dismissed, it once again came down to Brunel to find a better solution.

His unique design, which was a development of his Chepstow Railway Bridge design, which was used to cross the River Wye, consisted of dramatic lens-shaped trusses that linked two main spans with curved approach spans at each river bank. With his latest creation planned to sit 30 metres above the river, and with only one central pier, river access would not be significantly hindered, despite a monumental structure looming above the harbour at Saltash.

The elliptical forms on both spans created by the thick iron trusses above and gently curved chains below came as a result of negating any lateral force on the central pier. Unbeknownst to Brunel, this feature would in fact save the bridge from destruction during the Second World War, when a well-placed bomb from German forces was reportedly deflected by one of the metal arcs into the river below.

Brunel's achievements have not gone unnoticed in Saltash, with one public house proudly bearing his name. Note the depiction of one of the Royal Albert Bridge's spans above the sign. (Author)

Initial river surveys began in 1848 and by 1855 the iron trusses began to take shape, moored up against the Devon side's riverbank. Material was taken from Brunel's other bridge at Clifton, which had been halted pending funds and overdue deadlines, and by the winter of 1858 both Devonian and Cornish spans had been elevated to their impressive position above the riverbed. After final works had been completed six months later, inspections were made by train, and Prince Albert, who had agreed to have the bridge named in his honour, officially declared it open on 2 May 1859. Brunel's accelerating nephritis, a result of heavy smoking, ruled out an appearance for the ceremony, and perhaps added a physical challenge to the engineering difficulties faced in taking his railway to south-west England.

His efforts were greatly applauded however, with travellers and rail users up and down the country welcoming the railway's first steps into Cornwall. Locals declared satisfaction for being united with the rest of England, in equal measure to those elsewhere rejoicing the ease of exploring the country's southernmost tip. In addition, the Royal Albert Bridge became an instant Saltash landmark for both its grand stature and striking appearance, with Brunel too finding fame in the form of the town's many references to his success. Though the journey south posed a number of hurdles for Brunel and his Great Western Railway, by bringing mass transit to Cornwall he not only expanded his network, but also his fanbase.

The Tamar Bridge, built in 1961, now looms over the railway bridge, perhaps signifying the superiority of road transport in the modern age. (Author)

Chapter 21

Clifton Suspension Bridge, Bristol

Disruption was the theme for attempts at bridging the Avon at Bristol between the 1790s and 1860s. Riots erupted after the controversial implementation of tolls across the recently altered Bristol Bridge, resulting in eleven deaths, with many more injured. When plans were drawn up to span the gorge west of the city centre, a series of interruptions and interventions ensued, dragging out the creation of the resulting Clifton Suspension Bridge over many decades.

Initial plans ranged from elaborate stone viaducts stacked five layers high to simpler iron structures overhanging the vast valley between Clifton and Leigh Woods. Budget restrictions and a lack of agreement on specification saw a competition arranged in 1829, with the prize being the contract to build the bridge. Sarah Guppy, a Birmingham-born inventor, had contributed designs for a suspension bridge across the Avon to Isambard Kingdom Brunel, in doing so demonstrating her ingenuity and modesty. Guppy felt that, as a woman, presenting her proposals explicitly as her own may have been received

The impressively high cliff faces along this stretch of the Avon allowed for a number of designs to be put forward, without impeding the waterway below. (Author's Collection)

as boastful, and she described the notion of talking about oneself as 'unpleasant'. In addition to aiding Brunel, Guppy also provided free use of her patent for the foundations of suspension bridges to Thomas Telford.

Both Brunel and Telford put forward entries for the Avon bridge competition, Brunel as a participant and Telford as a judge, after rejecting all of the proposals by twenty-two designers and architects. A second competition was later won by Smith & Hawkes of Birmingham, but back-door persuasion ensured the contract was awarded to Brunel. Construction began in the summer of 1831, and soon after the people of Bristol again took to the streets, this time in protest over the House of Lords' decision to reject a bill giving political power to Britain's industrious cities. The riots halted progress on the Clifton Suspension Bridge and five years elapsed before construction resumed. Even then, the flow of cash to the project caused further delays, and with finances all but exhausted by 1843, some of the bridge's physical resources, such as the iron chains, were diverted to the Royal Albert Bridge in Cornwall – another Brunel endeavour.

It was only when Brunel died in 1859 that his admirers gathered further investment to complete the structure. Recycled materials from the Hungerford Bridge in London and a revised layout by William Henry Barlow and Sir John Hawkshaw ensured the stability of the bridge, which was opened with a ceremonial parade in 1864.

Despite battling many hurdles along the way, the iconic West Country landmark now acts as a monument to Bristol's position as a bustling city of the South West, as well as an example of the finest work by a number of Victorian engineers. In addition, Bristol now markets itself as a historical centre of industry for the south, particularly from a tourism perspective. The Clifton Suspension Bridge, complete with visitor centre, tour guides and explorable vaults (discovered only recently in 2002), adds to other major attractions such as the SS *Great Britain* and the Bristol Harbour Railway in allowing visitors to experience first-hand a number of historically important engineering feats.

Above: *This image highlights the advantages for shipping with regards to the bridge's impressive position above the Avon. (Author's Collection)*

Right: *The Clifton Suspension Bridge still sees frequent use as part of the UK road network, carrying the B3129 over the river via a toll. (Author)*

Chapter 22

Clifton Hampden Bridge

The nineteenth-century modernisation of Clifton Hampden, a small settlement located south of Oxford, was brought to fruition by the then director of the Bank of England, Henry Hucks Gibbs, who took control of the local manor in 1842. His amendments to the village largely centred on the creation of a new manor house, built on the northern bank of the River Thames, but also extended to rebuilding the local church before turning his attention to a river crossing.

Until then, crossing the Thames was done via the Clifton Ferry, a service owned by Oxford's Exeter College since 1493, or by wading through the adjacent ford. Once Gibbs had secured the service from the college and put forward his case for improving access to the area to parliament, he set on having the bridge constructed using bricks from Clifton Heath, which he had previously used for his other projects. Gibbs' motivation for constructing the bridge was arguably for personal gain, having found frustration in his staff, travelling from nearby Long Wittenham, missing the ferry and turning up late.

He is reported to have approached Sir George Gilbert Scott to design the bridge at the local Barley Mow pub where, in response, Scott was said to have outlined a sketch for a seven-arch bridge on the cuff of his shirt. Bridges were in fact not something Scott had

A fascinating quality of the River Thames is its varying scenery. Here, a horse peacefully takes water with the village church behind – a stark contrast to London's humdrum of activity. (Author's Collection)

been involved with before, having found notoriety for his wide array of churches, halls and school buildings. Once underway, his design was revised to only six arches, one of which was built to anticipate flood water on the eastern bank. Construction, courtesy of Richard Casey, was complete in 1867, following which Casey turned to keeping the toll house with his wife. While the new Clifton Hampden Bridge was intended to charge a toll, like almost all such crossings of the time, the fee was put in place merely to cover the costs of upkeep and not the total cost for building it, which was fronted by Gibbs.

In subsequent decades, the bridge gained praise for its design, colour and position on a stretch of Thames frequented by leisure craft. The release of Jerome K. Jerome's *Three Men in a Boat* in 1889 highlighted Clifton Hampden, and the Barley Mow in particular, as a place to moor up and spend the night – a notion still relevant for today's visitors. The introduction of motorised road traffic created an element of congestion for the bridge, and at one stage a replacement was in the offing, but following the conclusion of the Second World War and movement of ownership to the local council it was decided to keep the bridge intact, with it later being Grade II listed.

Above: *The Barley Mow, Clifton Hampden. (Author)*

Right: *From the northern bank, the attractive red hue of the locally sourced bricks can be appreciated. (Author)*

Chapter 23

Swing Bridge, Newcastle

Though the enormous feat of giving both road and rail simultaneous access across the Tyne had been achieved in the form of the High Level Bridge by 1849, it became apparent shortly afterwards that further access at a point closer to the river would be beneficial for the city of Newcastle. This access was in fact in place with the second major iteration of the Old Tyne Bridge, itself occupying the position of the Roman Pons Aelius. However, with river traffic being crucial for trade and transport, any improved road access would need to avoid disrupting ships and barges.

The Old Tyne Bridge was therefore demolished in 1868, but with the intention of replacing it with something more radical that, like the High Level Bridge, would satisfy two modes of transport around the river. William Armstrong, recently knighted for his success in artillery and willingness to offer patents to the government, saw the opportunity to further his business by having guns fitted to warships at Elswick, as well

TYNE BRIDGE, TAKEN DOWN 1866–73.

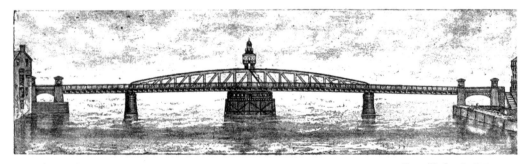

NEW HYDRAULIC SWING BRIDGE, OPENED 1876.

These diagrams depict the Swing Bridge and the Old Tyne Bridge that it replaced. (Author's Collection)

as providing something iconic for the city of his birth. His proposal was therefore for a swing bridge, which would have a central pivot and rotate 90 degrees to sit parallel to the river when open to water traffic, and perpendicular when closed.

Work began in 1873, entirely financed by Armstrong, and resulted in a wrought-iron bridge, with the rotating section being 86 metres long. Steam pumps were installed to enable the bridge to swing a full 360 degrees and was controlled by mechanisms held in a cupola at its centre. Once complete in 1876, much larger vessels were able to transit the Tyne, vastly increasing associated industries as well as giving access back to road traffic.

The two jetties stemming from the central and southern piers were integrated into the use of the bridge. The larger jetty, stretching upstream as far as the central pier of the High Level Bridge, at one time collected tolls from passing ships by employing someone to cast a large pole out to vessels with a small bag on the end. The smaller of the two, now slowly decaying into the river, once sat opposite the offices for the River Tyne Police. The police would moor apprehended vessels against the pier, giving the wooden structure its name, the Police Jetty. The offices opposite eventually became the River Beat Restaurant in recognition of its previous use.

Though the method of movement was converted from steam to electric in 1959, the mechanisms within the bridge remain as they were, and even with the post-war loss of most of the armament works at Elswick, the increase in road traffic ensured relevance through to the present day. Any large vessels still wishing to travel upstream now have to sail through two pivoting bridges, the latter being the Gateshead Millennium Bridge, though its method of movement is somewhat different to that of its older counterpart.

The High Level Bridge, Swing Bridge and Tyne Bridge together set an iconic scene for Newcastle. All three were featured digitally in a hidden level for the 1999 video game Driver. (Author)

Chapter 24

Axmouth Old Bridge

Concrete is now a staple building material in the construction of bridges, and though arguably overused at points in the mid-twentieth century, it still offers flexibility in design, placement and budget. Some sources even claim concrete to be the most utilised man-made material in history, and while it may be associated with the large Brutalist structures of the 1960s and '70s, concrete's origins go as far back as 6500 BC.

Two main types of concrete construction are used when developing any kind of meaningful structure: mass concreting, which simply uses the solid, compacted mass of the material to hold a rigid form, and reinforced concrete, which uses the placement of bars (most commonly metal) within the mass prior to setting to further stabilise the structure. As concrete construction became more popular throughout the nineteenth and twentieth centuries, and with the qualities of the material becoming more thoroughly understood, reinforced concrete became the favoured method.

Mass concreting took hold in France during the 1800s thanks to the efforts of architects such as François Cointeraux, and the technique was infrequently used in England, being employed in the construction of large dry docks such as Chatham and those on the River Clyde. Other more conventional buildings and bridges of the time were commonly made from brick or wrought iron.

Philip Brannon, an engineer, writer and artist, decided to use concrete when designing Axmouth Bridge, despite a number of failed construction projects on his native Isle of Wight and in London. The latter used the reinforced concrete method to create a number of buildings in Islington, only for them to collapse in 1874 due to poor construction.

Brannon's Devonshire bridge was to span the mouth of the River Axe as it met the shingle beach of Seaton in East Devon. As with many projects in the mid-nineteenth century, funding came courtesy of the expanding railway network and it was the Seaton & Beer Railway that saw value in facilitating access at a time when the area was exporting Beer stone from Seaton's railway station next to the river. Axmouth Bridge was opened in 1877 and a toll was implemented for those wishing to use it. At the time, the bridge was reportedly the third to be built using the mass concrete method.

Unlike the previous two, however, the bridge at the end of the Axe avoided demolition, but only just. Explosives were placed in preparation for the arrival of Hitler's forces during the Second World War, though thankfully they were never needed, and while the railway company, station and large-scale exportation of Beer stone have since disappeared as part of Seaton's economic decline, Brannon's concrete triumph stands as firm as ever. With the post-war surge in vehicle owners, the bridge eventually became insufficient for demand and a second bridge was built to suit, this time using reinforced concrete and opening in 1990.

Axmouth's Old Bridge, as it is now known, entertains foot passengers and cyclists looking to admire the small harbour located within the clutches of Seaton's prominent shingle spit. With the area's focus now firmly on tourism, modernised paving stones and lamp posts have tastefully enhanced the bridge's character, giving the illusion of a construction date much earlier than 1877. In fact, during the design

Right: *When viewed up close, the more ornate elements of the structure are revealed to be a result of the moulded concrete body. (Author)*

Below: *When viewed from afar, imperfections in the bridge's stance become apparent, likely owing to its age, but possibly also a sign of Brannon's pioneering but immature techniques. (Author)*

The toll house at the Seaton end of the bridge is immaculately kept. Until 1907, 1d was the charge for pedestrians, with 4d being the charge for horse and cart. (Author)

phase, Brannon made efforts to make the bridge look 'older', or certainly as if made from more traditional stone and not his innovative but bland concrete. Further adding to the shortcomings of appearance, the position of the newer adjacent bridge looms over its predecessor and cruelly obstructs the view up the Axe Estuary, which can provide many photogenic scenes during summer and is home to a wide array of river-based wildlife.

Those making the walk from Seaton's town centre up to Axmouth village (almost a mile upstream from the harbour) likely now cross the modern bridge without much consideration for the notable article beside them. Still, even with an underrated stance, the structure remains the oldest standing mass concrete bridge in England and signifies a landmark at the very forefront of a construction method that is still used frequently throughout the world.

Chapter 25

Forth Bridge

The most astonishing aspect of the Forth Bridge – more commonly referred to as the Forth Rail Bridge to distinguish itself from the later road bridge – is the network of metal that creates each of the three cantilevers spanning the Firth of Forth. Unlike a number of large structures built in the late 1800s, which made use of mass-produced iron, the Forth Bridge utilised steel and was one of the first prominent structures to do so. After the final design by John Fowler and Benjamin Baker was ratified, a large quantity of steel was produced from Landore, near Swansea, to provide beams and girders for the cantilevers, with the rest coming from a more local source at Glasgow, where some 4,200 rivets were also made. While Bessemer steel was being commonly produced at the time, it was mild steel made with the open-hearth Siemens Martin process that was used for this bridge as it was deemed to be of a higher quality and more cost effective.

Construction began in 1883, but prior to this preparations were being made at North and South Queensferry and on the small island of Inchgarvie, where buildings from the early 1500s lay in ruin. These buildings proved useful as offices, workshops and coastal defences for construction workers, in addition to newly placed facilities on both sides of the Forth. For a time, the surrounding area was transformed to allow the bridge to be built and became a landscape filled with hoisted material, labouring workers and the ever growing metal framework above the water.

The primary use for the steel struts were to cross-brace each of the three main structures, which themselves demonstrated, by virtue of their design, the principle of cantilevered spans. This principle was seen in basic form with the construction of stone bridges, with a few examples in metal emerging just prior to the construction of the Forth Bridge. Barker famously demonstrated his engineering by suspending his Japanese contemporary, Kaichi Watanabe, between the cantilever arms of two other men, held in tension. The opposing arms, stretched outwards away from the central span, acted as

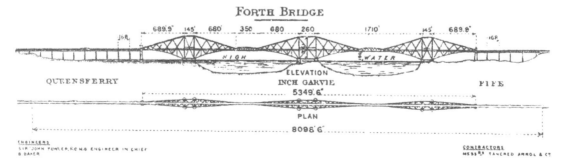

Plans showing the eastern side of the bridge and profile from above. (Author's Collection)

59

A haunting view of the cantilever's interior, showing the impressive network of steel throughout. (Author's Collection)

the anchor arms (also under tension), while wooden beams beneath them countered the force by being placed in compression.

Such was the scale of the superstructure, spanning the 8,000 feet across the Forth, that building it took seven years and required between 53,000 and 55,000 tons of steel, including the approach viaducts at either end. Some thought of the project as over-engineered following the Tay Bridge disaster, and though the bridge was another magnificent feat for the time, it was not without its critics. However, like Shropshire's Iron Bridge before it, the Forth Bridge excelled in the use of a newly abundant metal in a large-scale setting. When the bridge was opened in 1890, it was hailed not only as a success for its constructors, Sir William Arrol & Co., who would go on to build a number of iconic bridges, but also for Scotland as an example of world-leading engineering.

The resulting link for Scotland's railways greatly increased accessibility for settlements north of the Firth of Forth. Extra branch lines and further routes were laid purely to make the most of the bridge, which even today hosts local services to Edinburgh's suburbs and cross-country services to as far south as Penzance. In 2015 the Forth Bridge's monumental status (both in size and in historical importance) was recognised by UNESCO, who then inscribed it as a World Heritage Site, citing its forthright display of functional elements and industrial aesthetic as reasons to have it celebrated. Today, the Forth Bridge places Scotland firmly on the map, not only from an engineering perspective, but also as a key Scottish landmark. Many flock to South Queensferry's shore to admire the monumental structure, which is now accompanied by two further examples to the west. As a result, the means to better connect Scotland's cities to its Highlands has become one of the world's most widely recognised bridges.

An impressive scene in 1890, just as the final elements of construction were coming together. (Author's Collection)

An iconic image of Scotland, with floodlights tastefully adding to the celebrity status of a world-renowned engineering feat. (Author)

Chapter 26

Tower Bridge

When an Act of Parliament was passed in August 1885 to allow for the construction of a bridge across the Thames at London's East End, concerns were raised about accommodating the needs of not only pedestrians and road users, but also those of the Pool of London and the countless vessels required to dock within. While great care was taken to ensure that the aesthetics were more than pleasing (as was the norm during the Industrial Revolution), even those involved in the concept of Tower Bridge could not have anticipated the extent to which, over 100 years later, the bridge is now admired as a worldwide icon.

As with almost every bridge crossing the 215-mile-long River Thames, there was a desire to connect opposing sides to ease the flow of people and goods in England's capital, despite London Bridge (predecessor to the current one of 1973) lying half a mile upstream. The vast increase of industry and trade, along with an ever-growing population, made it necessary for multiple spans to appear along the length of the waterway.

After receiving royal assent stipulating particular details in the design, the job was awarded to John Wolfe Barry, with several other concepts, including one by Joseph Bazalgette of sewerage fame, being rejected by a panel of judges. Specific attention was drawn to water traffic so that it would not be hindered by the addition of another bridge, thus working in favour of Barry's bascule design, which allowed for the required 200 metres of clearance for vessels passing through. A Gothic style was also a requirement of any proposed plan.

Construction took eight years and made use of up to five prominent contractors. Over 400 workers saw the rise first of the foundations and piers, followed by the two bridge towers, the engine room on the river's south side and finally the ornate facade. This was altered to a Victorian Gothic style, in line with the original guidelines, by George Stephenson after the death of architect Horace Jones in 1886. Jones' original plans for the external aspects of the bridge involved brick, but it was ultimately Portland stone that finished off the completed structure's grand appearance.

On 30 June 1894, Tower Bridge was opened with all the pageantry and fanfare expected for a monument of its calibre. With the towers topped at 65 metres, the bridge dwarfed a large number of buildings on the riverside and complemented the much older Tower of London on its north side. It was from here that a salute from ten Crimean War-era guns marked the occasion and, on behalf of Queen Victoria, the Prince of Wales (the future King Edward VII) officially declared London's latest span open, exercising his desire to become a more contemporary and publicly visible royal.

In time the bridge gained fame for being one of the most striking of London's landmarks and soon became a flagship symbol for the British Empire. After avoiding the Luftwaffe during the Second World War and being granted substantial modernisation in

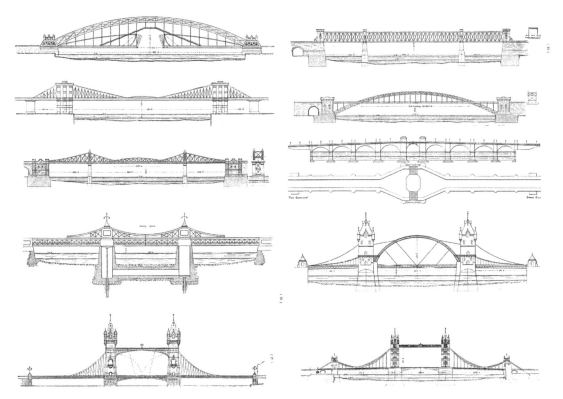

Above: *A number of proposals showed the differing philosophies of prominent engineers and architects, including Joseph Bazalgette, Horace Jones, John Wolfe Barry and A. J. Sedley. (Author's Collection)*

Right: *Beneath the ornate stonework lies a staggering quantity of iron and steel, as depicted here. (Author's Collection)*

the 1970s, focus shifted to making use of Tower Bridge's celebrity status in the tourism industry. The importance of Tower Bridge's practical function had diminished with the completion of London Bridge's third reincarnation, and even if its novel bascule feature still meant much to the river's larger visiting vessels, road traffic now had the Vauxhall, Lambeth, Southwark and Waterloo bridges easing congestion.

Today the landmark is partly responsible for bringing over fifteen million tourists to London each year and is seen on almost every postcard, poster and promotional leaflet. This influence in the twenty-first century spreads far across the world, to the point where China, while seemingly replicating the success of Britain's Industrial Revolution, had its own version of the world-renowned landmark built. In a bizarre display of one-upmanship, Suzhou's 'Tower Bridge' hosts four towers and as many walkways constructed with almost every detail present on Barry, Jones and Stephenson's original designs.

Whether this gesture signifies flattery or fraudulence remains to be determined, but either way it is clear that Tower Bridge holds a legacy for Britain far beyond its original purpose as just another Thames crossing.

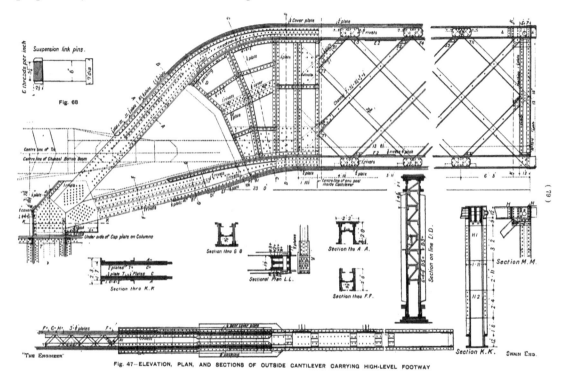

Fig. 47—ELEVATION, PLAN, AND SECTIONS OF OUTSIDE CANTILEVER CARRYING HIGH-LEVEL FOOTWAY

Above: *The walkways now contain the Tower Bridge Exhibition – a far cry from their previous use as a hotspot for prostitutes. (Author's Collection)*

Left: *As if to go one better in the Far East, Suzhou in China has 'doubled' the original Tower Bridge to create an over-exuberant highway and photo opportunity for visitors. (Alessio Avezzano)*

Chapter 27

North Bridge, Edinburgh

Edinburgh's unique geographical features posed particular problems for the historic town during the eighteenth century. The steady rise in population resulted in the rocky outcrop, headed by the famous castle, reaching its breaking point along the Royal Mile, with seemingly no option to extend the city outwards. The primary barrier was the valley to the north, where over time a body of water known as Nor Loch was situated. A rise in land that extended downward towards the Firth of Forth provided ample room for Edinburgh's population to grow, and so, after some debate and much promotion by the Lord Provost, George Drummond, it was decided to bridge the valley.

William Mylne, whose brother Robert had already bridged the Thames and Isle of Seil, was appointed architect for North Bridge and set about creating a sloping structure of five arches. His work was in parallel with the draining of Nor Loch and the creation of a large public garden in its place, but loose material from previous building efforts contained within the valley proved unstable for the predicted depth required for the bridge's foundations. In August 1769, five people were killed when a section of the structure collapsed, and though repairs were costly, the bridge eventually opened in 1772. Future landscaping would go on to make use of the discarded loose earth in the form of The Mound, which was completed in 1830 and located to the west of the bridge.

Ambitious plans to extend the growing railway network through Edinburgh began to gather pace shortly afterwards, and though great effort had been made to beautify the valley between the old and new towns, by 1884 parliament gave permission for a path to be made through the gardens for the construction of a railway line and station. This drastic revision of the valley led to the necessity of a replacement bridge and, off the back of their applauded efforts with the Forth Bridge, contractors Sir William Arol & Co. began to construct a second North Bridge in place of Mylne's. Unlike the previous

The first stone bridge cast a shadow over the large public space below before it made way for the railways. (Author's Collection)

bridge, which was entirely of stone, six arched girders of iron were placed atop stone piers, with exterior decoration coming in the form of cast iron, which were designed by Robert Moreham. After three years of construction, which involved some buildings of the old town opening new entrances onto the roadside, as well as the implementation of railway infrastructure, the new North Bridge was opened on 15 September 1897.

The work was carried out together with that of Waverley Bridge, which even today forms part of the railway station below and facilitates access via two roadways connected directly to platform level. Though original proposals sought to limit the visual impact of the railway in and around Princes Street Gardens, from either bridge the view is dominated by the large canopies, numerous tracks and the audible movement of trains. George Drummond's desire to extend the city, though not via the bridge he laid the first stone for, was greatly fulfilled, with the present North Bridge now connecting two hugely important parts of Scotland's capital.

Above: *Prior to the building of the large canopies over the many platforms at Waverley station, the hive of activity below could be seen by the people of Edinburgh. (Author's Collection)*

Left: *To compensate for its intimidating stance over the city, the present North Bridge is beautifully decorated. Former railway hotel The Balmoral sits at one end. (Author)*

Chapter 28

Newport Transporter Bridge

Over many centuries, a variety of bridge building styles arose from the restrictions put in place by various maritime authorities. As significant bridging efforts largely spanned bodies of water, requirements to keep Britain's naval and merchant interests flowing were sternly enforced, often giving architects and engineers an additional challenge to those proposed by the physical environment. At the end of the nineteenth century, Newport in South Wales was making the most of a monumental coal industry. Though nearby Cardiff had since dominated the scale of operations, it was Newport that had once led the industrialisation of Welsh coal with large docks south of the town's centre.

To better connect people to their workplace, the River Usk required a span of its narrow meanders that did not hinder the activity beneath. Europe had already capitalised on a design snubbed in the 1870s by Glasgow and Hartlepool that featured a steel framework built high over the river, hosting a gondola that would swing between either bank, transporting passengers and goods in the process. By 1900, examples were

The gondola hangs over the River Usk during a period of closure. Today, visitors have the chance to cross the upper gantry instead. (Alan Reynolds)

erected in France and Spain, and after dismissing the high costs of building a tunnel beneath the river, local authorities decided to apply the same bridging concept in Newport, using a design by Frenchman Ferdinand Arnodin.

The structure was completed in 1906, with two electric motors driving a platform from bank to bank. Though Cardiff had taken much of Newport's coal-related activity away, the town's diversification allowed the docks to make use of other industries in steel, iron and engineering. However, the town subsequently began to decline and with the rise of the motor car, Newport's Transporter bridge quickly became insufficient for Britain's road traffic, but remarkably held its own as part of the road network until closure in 1985. Following this, the transporter bridge concept arguably gained a new figurehead at Middlesbrough, where the Tees Transporter Bridge, opened in 1911, had operated more significantly as part of the road network, forming a section of the A178.

Newport's bridge now serves as a heritage icon and has received notable funding since the settlement was granted city status in 2002. Renovation allows the gondola to carry tourists and motorists for a small fee and the upper gantry can be accessed by visitors wanting to capture the view of Newport's bygone industrial landscape. With only a small number of such structures still left in the world, the Transporter Bridge now demonstrates the long-term appeal of what could arguably be described as a niche bridging concept.

Silhouetted against the sky, the maze of metal work resembles a contemporary electricity pylon. (Author)

Chapter 29

Hertford Bridge, Oxford

Such is the vast history of the University of Oxford that few people go as far as marking a specific year in which the academic institution was founded. Tuition most likely started during the eleventh and twelfth centuries, and since then Oxford has gone on to host one of the most renowned universities in the world. Equally as old as the university are some of the buildings that line the city's streets. Somewhat ironically, New College, though founded a little time later, in 1379, lies within some of Oxford's oldest buildings and over the subsequent centuries, numerous colleges were founded and accommodated via a vast complex of structures in a variety of building styles.

When Sir Thomas Graham Jackson came to making his own additions to the university's architectural legacy, he did so mostly within the confines of Hertford College, once known simply as Hart Hall. In addition to other projects at the Examination Halls, Brasenose College and Trinity College, he implemented a small covered bridge between two quadrangles of Hertford College, enabling the ease of passage for academics. The bridge, completed in 1914, featured two sets of stairs, forming an arch that, combined with its external appearance, closely resembled the Rialto Bridge in Venice.

However, over time the structure became known as the Bridge of Sighs, reportedly due to its similarity to the bridge in Venice of the same name. This bridge, while functionally

Despite its association with the Bridge of Sighs in Venice, Hertford Bridge has much more in common with the Rialto Bridge, shown here. (Author's Collection)

Above: *With Oxford a prime destination for overseas tourists, the bridge sees plenty of attention from photographers. It also previously featured on the cover of the area's Ordnance Survey Map. (Pauline Gower)*

Left: *The actual Bridge of Sighs looks distinctively different, thus complicating the origin of Hertford Bridge's alter-ego. (Author's Collection)*

comparable in that it connected two buildings (this time a prison and an interrogation room), had its own distinctive facade in a different style to that of Jackson's, and quite why or how the title was bestowed upon the Oxford span seems to this day to be a case of mistaking the two Venetian bridges. To add to the confusion, the moniker spread to a number of other locations in England, including Oxford's academic counterpart, Cambridge, whose own Bridge of Sighs is named directly from the Italian example.

Italian influence was certainly a factor in Jackson's image. Though the Clipsham stone had origins in Rutland and Caithness, the carvings found on its exterior were Jackson's own interpretation of the English Renaissance that he himself admitted was more popular in Italy and arguably out of place in Oxford.

Reportedly, the 'sighs' from which the Venetian bridge got its name came from convicts who, as they glanced through the bridge's barred windows, saw their last view of the outside world before being imprisoned. This concept carried over to Chester's Bridge of Sighs, notably basic in comparison to others, this time with convicts sighing in anticipation of execution.

Oxford too deciphered a deeper meaning for its new designation for Hertford Bridge and compared the need to sigh prior to death or imprisonment to that of preparing for inevitable exams. Though almost certainly a means to justify the name's puzzling origin, this notion adds to the mystifying qualities of life on one of the world's oldest campuses, particularly now that Oxford has established itself as one of England's foremost tourist destinations.

Chapter 30

Waterloo Bridge, London

A number of significant factors signalled the demolition of the first Waterloo Bridge during the 1930s. Though John Rennie's career had seen great achievements in Britain's canals, including features such as the Dundas Aqueduct near Bath, his ability to translate this to a large span across the Thames was eventually deemed unsuccessful. The bridge was at one time renowned for tragic suicides and accidents, taking the life of the American performer Samuel Gilbert Scott in 1841 after he fell from scaffolding, but it was soon noted for its structural weaknesses too. Despite this and the longevity of his work on Southwark Bridge and Old Vauxhall Bridge, his work on the first Waterloo Bridge was widely celebrated.

Work on a replacement began soon after the original was dismantled and saw guidance from Giles Gilbert Scott (no relation to the American stuntman), an architect who had already altered the landscape of Britain with his iconic red K2 telephone box. His use of Portland stone for the exterior, as with a number of high-profile London landmarks, ensured a uniform appearance across each of the five spans. Concrete was used for the main structure and piers, though some difficulty was found in accommodating its sleek, slender shape.

John Rennie's bridge was certainly better looking, but structurally not as sound. (Author's Collection)

The outbreak of the Second World War saw work on the bridge halted before being continued by a largely female workforce. The bridge was the only one on the Thames to be successfully targeted during early air raids, being hit on 10 May 1941, but work continued until it was complete in 1945. Through necessity, the bridge had in fact been officially opened in 1942 and was for a time referred to as the 'Ladies Bridge' in recognition of its labourers from contractor Peter Lind & Co. Ltd. Following its completion, it was noted that Herbert Morrisson, a former council leader, omitted this feature of the workforce in his speech, and it was only after further research that women's efforts on the bridge during the war were truly recognised.

Tragedy struck the new bridge once again in 1978, when Georgi Markov, a Bulgarian novelist and dissident, found himself at the mercy of a complex assassination, in which a ricin pellet was fired into his thigh through method of an elaborate umbrella gun fashioned by the Bulgarian Secret Police. In the following decades, the bridge has been at the centre of many culturally iconic moments, including the memorable ending to the 1996 film *Trainspotting*, Coldplay's music video for the single 'Fix You' and a number of romantic references in plays, songs and poems. Practically, the bridge now forms a crucial part of the road network in and around London Waterloo railway station and across the Thames to the Embankment and Covent Garden. Its position on a sharp meander also provides visitors with the ability to see upstream to the Houses of Parliament and the London Eye, as well as the Shard downstream and many other iconic bridges along the river.

Despite a detailed history and notable architect, the finished article is arguably rather plain and functional. (Author's Collection)

Above: *Sir Giles Gilbert Scott's involvement with Battersea Power Station, another prominent project, was during the latter stage of its development. (Author)*

Right: *The death of Georgi Markov made headline news in 1978. His grave lies in the quiet Dorset village of Whitchurch Canonicorum. (Author)*

Chapter 31

Kingsgate Bridge, Durham

When Ove Arup founded his firm Ove N. Arup Consulting Engineers in 1946, he may not have anticipated the extent to which it would influence some of the world's most recognised engineering feats. His initial motivation stemmed from the desire to involve multiple disciplines in the creative process for construction, rather that have them develop individually. This resulted in achievements so substantial during the twentieth century that it is a wonder that his name is not more frequently raised in conversation.

His involvement with D-Day's Mulberry Harbours was crucial, enabling the militarisation of the north French coast and aiding the Allies' invasion of Normandy. He also consulted the government on the subject of bomb shelters, in both cases promoting the mass use of concrete. Following the war, his firm, simply referred to as Arup, reincarnated the remains of Coventry Cathedral, which was severely damaged by the Luftwaffe, into a public art space, as well as building the new cathedral beside it. Though the strikingly modern design split opinion at the time, it has since been embraced by the city.

Durham's position on The Bailey means that a number of bridges traverse the River Wear surrounding it. The oldest is Framwellgate Bridge. (Author's Collection)

In 1963, Arup made the decision to personally supervise the design and construction of the Kingsgate Bridge in Durham – a contribution that seemed minor in comparison to his firm's wartime building efforts. He devoted himself to every aspect of the structure, highlighting his affection for the area of his birth, despite his strong Danish heritage. The bridge, which connected The Bailey peninsula to Dunelm House, marked a first for Arup and featured a thin walkway supported by two sharply angled legs stemming from bases either side of the River Wear. As with notable previous works, reinforced concrete was used to create the span, which was built in two halves before being joined in the middle by expansion points made of bronze.

Though the firm's ambitions soared to dazzling heights after the Kingsgate Bridge was finished, it stood arguably as a testament to Arup's dedication to his field. His death in 1988 came prior to the construction of many iconic landmarks, including London's own Millennium Bridge, the 'Bird's Nest' National Stadium in Beijing and the corporate headquarters of Apple in California. Even though he had seen the completion of the Sydney Opera House in 1973, which can arguably be considered an engineering centrepiece, such was the importance of Durham's modest footbridge to Arup that it became the place from which his ashes were scattered.

Kingsgate Bridge today holds a number of awards and accolades, some in recognition of its unique design, others to celebrate the outstanding work of one of the world's leading engineering firms and its passionate founder.

Above left: *Kingsgate Bridge is an unusual, but striking addition to Durham's historic architectural landscape. (Author)*

Above right: *The 1970s saw Arup construct their most notable structure to date – the Sydney Opera House. (James Jones)*

Chapter 32

Severn Bridge

By the 1890s the railways had already seen the implementation of infrastructure to allow trains to cross the River Severn in the form of two major construction projects. The Great Western Railway, adding to its feats of the Box Tunnel and the Wharncliffe Viaduct, furthered the line west of Bristol beneath the river near Pilning to create the Severn Tunnel – the world's longest underwater tunnel until 1987. At the same time, with the expansion of coal mining in the Forest of Dean, a bridge was constructed at Lydney by the Severn Bridge Railway to gain easier access to Sharpness and beyond.

In comparison, by the mid-twentieth century, road access across the river was yet to be fully realised. Over Bridge, completed by Thomas Telford in 1828, allowed crossing by road near Gloucester for some time, but with the vast increase in motor vehicles, a more ambitious solution was required. A national road network was envisaged by the government following the Second World War and plans were drawn up to implement a road crossing downstream at Chepstow. Construction began in 1961, three years after work began on the Forth Road Bridge, which was born from the same scheme.

Above: *Physically, the bridge links Aust and Beachley, both in England, though traffic is then carried into Wales via the Beachley Viaduct and Wye Bridge. (Author)*

Left: *Generous access for cyclists and pedestrians on the south side of the Severn Bridge makes for a great run between England and Wales. (Karen Rogers)*

This delay, however, saw the Severn Bridge make use of an innovative road deck design. Dr William Brown concluded that a lightweight, hollow structure would not only be easier to build, but would also deflect crosswinds much better than the heavier deck portions used for its Scottish counterpart, which itself was based on a number of bridges in the United States. In total, eighty-eight sections were used, being hoisted into place above the water to form part of the M4 motorway (later renumbered M48).

By this time, road traffic was beginning to have a serious impact on the railway network. The Great Western Railway had originally protested the idea for a road crossing, predicting the inevitable change from the mass movement of people to personalised transport in the form of cars. In addition, coal mining was dwindling, and after a number of high-profile incidents, the Severn Railway Bridge at Lydney was demolished, rather fittingly with the help of *Severn King*, a vessel left with little purpose after its previous life as the Aust Ferry was made obsolete by the new road bridge.

Queen Elizabeth II opened the bridge on 8 September 1966, marking the beginning for a new flow of traffic to the south of Wales. Much like the Forth Road Bridge, the need for a replacement became apparent during the 1980s, and over the next twenty years, a series of measures mitigating the corrosion of the cables were put in place. Restrictions began during this time, and during the 1990s the Second Severn Crossing (later renamed the Prince of Wales Bridge) materialised to take on the burden of heavier use, as well as simultaneously spanning the river even further downstream from its predecessor.

Above: *The Second Severn Crossing, renamed the Prince of Wales Bridge in 2018, has two dramatically curved viaducts leading to the central span. (Author)*

Right: *From the same angle, the old Severn Bridge offers little in the way of curves. (Author)*

Chapter 33

Kingston Bridge, Glasgow

As a city bearing the brunt of Scotland's industrial and economic activity, over time Glasgow has required major overhauls to its busy transport infrastructure. With the growth of shipbuilding in particular, the River Clyde became the centre for the city's labour, and as the population increased to suit, the practicalities of moving workers around became ever more ambitious.

Following the opening of the London Underground in 1863, Glasgow too decided to dig its way to mass transit, and by 1896 the uniquely designed Glasgow Subway was shuttling passengers across the city and underneath the river. Above ground, a number of bridges had been built to further link the north and south sides of the Clyde,

The South Portland Street Suspension Bridge seems delicate in contrast to its companion downstream. (Author)

including the South Portland Suspension Bridge, which, from 1853, enabled access to the city centre for the poor labourers of the Gorbals. Nowadays, it serves as a somewhat neglected pedestrian footbridge.

By the 1960s, Britain's road revolution resulted in the mass building of nationwide motorways. In tandem with a scheme to redevelop Glasgow and to facilitate access to the rest of Scotland, the M8 motorway was planned to cross the Clyde above the disused Kingston Docks. Despite hosting the city's first enclosed dock and a large number of historic buildings, the entire area became a scene of unrelenting construction as the immense ten-lane roadway took shape, curving through Anderston before being elevated to a height of 18 metres across the river. The bridge marked just one section of a massively ambitious scheme to encircle Glasgow with roadways, but it soon fell short of the city's need to accommodate Britain's motorists.

By the time of its opening in 1970, the imposing concrete structure was thought to be ready to bear the weight of Glasgow's motor car masses, but plans to seamlessly integrate a major motorway system into the city had stumbled in a number of key areas. In addition to the embarrassment of a stillborn section of highway existing immediately south-east of the bridge, it became clear that the Kingston Bridge would not cope with the sheer volume of traffic, which by 1990 regularly exceeded the predicted figure of 120,000 vehicles per day. As a result, major restructuring took place to redo what was seen to be a flawed piece of engineering and restrictions to traffic were imposed in 1994. Such was the scale of the plans to strengthen the bridge that *The Guinness Book*

Though part of a larger failure to impose the car on Glasgow, Kingston Bridge now successfully carries ten lanes of traffic via the M8 motorway over the Clyde. (Author)

of Records declared it to be the world's largest ever bridge lift when 128 hydraulic jacks lifted the structure onto newly built supports. The bridge had already earned an accolade as Europe's busiest bridge.

Over time, in contrast to the overwhelmingly negative position on its appearance and consistent closure, the bridge did sometimes ignite Glaswegians' imaginations. Sinister rumours of gang-like burials within the concrete mirrored similar tales from across the pond following the death of Jimmy Hoffa, and the extent to which the urban highways of Glasgow could have inspired the free and passion-fuelled youth is arguably portrayed in the music video to Simple Minds' 1984 hit 'Speed Your Love to Me', which features much of the M8 infrastructure, including the Kingston Bridge, in a sequence of shots that marries fast-paced motoring with soaring guitar.

Today, with the regeneration of Glasgow's famed riverside the once-bold transport amenity, while still holding an impressive stature, looks aged. Modern apartments, retail parks and contemporary business centres try their best to involve the bridge in the city's landscape, while the oft-empty relics of industrial Glasgow remind onlookers of a time before the motor car. The Kingston Bridge signifies the impact that road users had on Glasgow, along with the aspirations of ambitious planners and architects, which altered the flow of Scotland's second city forever.

Cynics could justify the bridge's width as a means of sheltering pedestrians below from Glasgow's inclement weather. (Author)

Chapter 34

Itchen Bridge, Southampton

One of the earliest significant proposals for a bridge to cross the lower extremities of the River Itchen was made in 1885 by the newly formed Itchen Bridge Company. At this time, Woolston – named for its historical involvement in the wool trade – was in need of a link to Southampton, and after consideration for a permanent roadway subsided, it was left up to the Woolston Floating Bridge, a long-standing ferry service, to offer transport between the two settlements.

Renewed interest began once the town had been absorbed as a suburb of Southampton in the 1920s and with the increase in popularity of the motor car. The ferry service was expanded to allow vehicles to cross and ran regular services until the 1970s, when it was finally decided to implement a bridge in its place. The decision came as an unpopular one for locals, who regarded the Floating Bridge as part of Woolston's identity and the area's maritime history.

The government refused to provide funding for the project, which was effectively finalised during the 1960s, and so the local council was left to raise over £12 million to fund the bridge. By 1974 construction had begun on a substantial five-span concrete road bridge that dominated the local landscape and resulted in areas of housing east of the river being relocated.

'Southampton from the Water', *an 1850 depiction by Philip Brannon, architect of the Axmouth Bridge. (Author's Collection)*

Above: *Despite being despised by many, the Itchen Bridge carries itself proudly over the river. (Author)*

Left: *Certain viewpoints give the impression of a floating highway, well above the tired and patchy urban environment beneath. (Author)*

Piers were initially erected to support the building of the deck outwards from each to form a balanced cantilever design. The two outward spans were slightly shorter than the other three, but at approximately 800 metres long, the narrow profile and proportions of the bridge were sleek and striking. Its height of 28 metres allowed for large vessels to travel through the central span and out toward the Solent, while road traffic flowed overhead.

With the loss of the ferry, the rearrangement of Woolston's residents and its slender but grey overall appearance, the reception of the Itchen Bridge was largely negative when it was opened in 1977. Further issues rose from the fluctuating toll rates imposed to finance the bridge over subsequent decades, resulting in a complex automated system based on axle height to be install during the winter of 2010. Its poor level of maintenance, evidenced by a lack of safety features for pedestrians, clogged expansion points and frequent vandalism have given the structure an undesirable reputation for being a frequent location for crime. Estimates put the number of suicides committed from the bridge at over 200 since it opened and petitions to make alterations to the bridge have been known to gather momentum.

However, the Itchen Bridge does act as a notable landmark on Southampton's cityscape and carries a major thoroughfare in the form of the A3025 directly into the city centre. It is also the most prominent crossing of the Itchen and symbolises, somewhat controversially, the importance that crossing the river holds for the area.

Chapter 35

Humber Bridge

Crossing the River Humber was a task performed by some of the earliest civilised settlements in North East England. The ancient fort of Petuaria, built by the Romans at the site of present-day Brough, likely existed owing to its position at a narrow meander of the river where crossing was somewhat easier.

Even though the concept of a permanent bridge was available during this time, albeit in a basic format, the wide, mud-laden estuary with its varying tide and flow posed too much of a technical feat for the Romans, and certainly the Celts in the first century. Aside from wading through the water at low tide, an option taken up by some, the only alternative to travelling countless miles inland before returning to the other side of the river was by ferry. Though the earliest records of a Humber Ferry start in 1315, this method likely saw use prior to this at locations such as Brough and further downstream at Kingston upon Hull.

Following the end of the Second World War, the dawn of the motor car quickly placed pressure on the surrounding road infrastructure and the long-standing ferry, which was only equipped to transport a handful of vehicles. By the 1970s motorists in Hull had to travel as far inland as Goole along the M62, before travelling back toward the coast to reach the likes of Scunthorpe and Grimsby. Despite the 2,000-year time difference, the inconvenience was a commonality between the people of Humberside and the settlers from ancient history.

The two towers near completion. Even without the road deck between them, the bridge imposes itself on the Humber. (Crosby Stone)

With the cables in place, the first of the road deck sections can be seen suspended above the river. (Crosby Stone)

The Humber Bridge therefore became a welcome addition to the estuary when construction began in 1972. Its inception was the result of planning that started in the 1930s, and by the '60s it had become a crucial tool in gaining support for the then Prime Minister Harold Wilson, whose Labour Party arguably secured the 1966 Hull by-election by promising a span across the Humber. The suspension type design was the product of Bernard Wex, a former tank commander in the Army, and was destined to become the world's longest suspension bridge. Slip forming, a technique popular in the United States for building tall concrete structures, saw the erection of both towers before the cables (weighing 5,500 tons each) were placed and the sections of road deck hoisted into position. Both towers stood at 155.5 metres tall on completion, and were further from each other at the top, owing to the curvature of the earth.

Now in the routine of cutting the ribbon on an increasing number of impressive bridges in Britain, Queen Elizabeth II declared the Humber Bridge open on 17 July 1981, thus reducing the road distance across the Humber by up to 50 miles.

In 2017, the bridge was given Grade I listed status, both for its importance in engineering and also as an object of beauty. Words used in Historic England's reasons for designation include 'elegance', 'harmony' and 'intimately', which do not always associate themselves with large concrete products of the 1960s and '70s. Its ability to finally connect people from Hull to those in Lincolnshire's industrial towns was, and still is, widely celebrated, and though 'Humberside' has faded from being an officially recognised county, the principle of the River Humber's people being unified is still very much alive today.

The sheer size of the Humber Bridge alone marks it as one of the greatest structures in Britain, though at one point it stood globally at the forefront of bridge-building. (Author)

With the footpaths on either side lowered below road level, pedestrians are separated from the constant flow of traffic. (Author)

Chapter 36

Millennium Bridge, London

With our prolific use of roads and railways, hidden networks of utility pipes and historic use of canals, bridges in the modern age are often designed to accommodate a multitude of purposes. In some parts of the UK, the idea of building a bridge simply to carry pedestrians or cyclists might seem a waste when other disciplines of infrastructure might benefit from a concerted effort to span an obstacle.

The reality is, however, that the first bridges would have been built to simply facilitate crossing by foot. In the late 1990s, one of the primary objectives of the those involved with designing Millennium Bridge in Central London was to link Southwark and St Paul's purely through the flow of the capital's pedestrians to 'encourage new life on the embankment'.

However, as the second millennium came to a close, proposals for urban footbridges also found themselves at the heart of a drive to diversify local landscapes, make

Above left: *Despite an air of the contemporary 'form over function' philosophy, Millennium Bridge displays a number of its structural elements candidly to onlookers. (Author)*

Above right: *The Millennium Bridge carries pedestrians directly into the heart of the City of London. (Author)*

statements about the people and, at times, be outrageous with aesthetics. Artistic flair saw short spans appear in Bristol and Manchester, with Pero's Bridge at Bristol's floating harbour bypassing a walk of just over 500 metres, and the latter carrying Mancunian shoppers over Corporation Street. Though both of these projects did have a legitimate bridging purpose, the extent of their practical use came second to their part in a wider regeneration of each city for the year 2000.

Millennium Bridge too had strong artistic connotations. The sharp aluminium deck with shallow suspension cables resulted in a slender arc that aimed to allow uninterrupted views across one of the world's most iconic cityscapes. Not only was the bridge designed to look striking and modern, but it also tried to give London a similar vibe, making the project as much about the city as a whole as the mere link between the north and south banks of the river. Whether this intimate connection still held when the bridge was shut to reduce synchronous lateral excitation – otherwise known as wobbling – remains debatable, and certainly the press, along with cynical critics, were quick to highlight the possibility that the structure was an obvious case of form over function.

The bridge was finally reopened on 22 February 2002 and has since solidified its position as an iconic piece of London's vast infrastructure. Photographers intent on capturing Canary Wharf, the Houses of Parliament or Tower Bridge now use the bridge's metal platform as yet another medium to promote one of Europe's true World Cities. Despite this, commuters will tell you that aside from its architectural or cultural merit, Millennium Bridge still fulfils an underlying practical purpose in line with the very essence of bridging.

Though at one time London's latest addition, the city continues to build around the bridge. From Peter's Hill the river is not even visible. (Author)

Chapter 37

Salford Quays Millennium Lift Bridge, Salford

In the run up to the year 2000, Britain's series of investments into the regeneration of urban areas outside of London conjured a number of radical and thought-provoking architectural projects. This was largely prompted by a shift in industry, with such sectors as shipbuilding, fishing, car manufacturing and mining all seeing success overseas and leaving large areas of redundant infrastructure out in the open to decay. Manchester Docks, once the third busiest port in Britain, suffered from the limits imposed by the Manchester Ship Canal and was in a severe state of decline in the ten years prior to its closure in 1982.

The increase in the service and technology industries prompted a means to regenerate sites such as Salford during the 1990s, and after an injection of finances and optimism, work began on creating an area centred on not only these innovative sectors, but also the arts, the public and Manchester's cultural importance in the UK. The result was Salford Quays, which by 2011 had not only seen a host of metal-clad offices, apartments and public art features installed, but also The Lowry, an arts centre dedicated to the artist L. S. Lowry, and the boldly designed Imperial War Museum North.

Connecting these new areas of commerce and culture required a bridge across the canal to suit the aesthetic of regenerated Britain, while allowing vessels to sail through

Flash apartments, media facilities and glass-fronted office blocks provide company for the Lift Bridge – a stark contrast to the warehouses and shipyards of the past. (Author)

the area unhindered. Parkman of Manchester, at the helm of the bridge project, opted for a design in keeping with the modern millennium theme, featuring two joined arches on a deck that would be lifted by four steel towers containing concrete counterweights. After its completion in 2000, which involved the deck being slotted between the opposing pairs of towers, the bridge acted as one of the first major pieces of infrastructure to facilitate movement between the two areas of former dockland. A lift time of only three minutes, combined with artistic LED lighting, a striking silhouette and an orthotropic layout, ensured the bridge's status as being at the forefront of technology, and its adopted name as the Lowry Bridge integrated it into the historical context of Salford and Manchester.

With the creation of MediaCityUK, the new home for much of the BBC's operations, the area now sees significant use, dragging the area out from its depressing state symbolised poetically in The Smiths' back catalogue to one of the great successes of British redevelopment. Similar templates now applied to Cardiff, London, Birmingham and Liverpool have gone on to express a willingness of the UK to keep up with changing times and implement facilities to excel the country through the twenty-first century.

The most striking feature from this image is the matching of the deck's surface with Quay West's bronze cladding. (Karen Rogers)

Salford Quays extends Manchester's renewed urban landscape beyond the city centre. (Author)

Chapter 38

Steampipe Bridge, University of Birmingham

The vast majority of bridges have allowed for the movement of people in one form of another. Over the centuries this has grown from simple footbridges to vast railway and road structures, carrying hundreds of thousands of people per year. In making the most of the surrounding girders, framework and structural elements, it is not uncommon for utilities to be carried through a bridge, and a number of examples in the UK see electricity, sewage and gas piped across their underbelly.

In an area of Britain dominated by utilitarian structures, the Steampipe Bridge seeks to bring an artistic flair to the purposeful process of heat transfer. (Author)

However, the decision to create a standalone utility bridge is somewhat unusual. Often this requires spanning short distances near to power stations, sewage works or refineries and consists simply of pipework extended over a river or valley. Following the implementation of a sustainable energy strategy at the University of Birmingham, which involved the use of Combined Heat and Power (or CHP), it became apparent that heat would need to be transferred across the Worcester & Birmingham Canal and the Cross-City Railway line, which runs parallel.

Given the existence of a road bridge crossing University railway station and the tight confines surrounding the Victorian canal and railway, MJP Architects were called in to design a utility bridge that could carry waste heat from the energy generation plant to the medical school. A number of factors complicated the proposal including the distance above the electrified railway and canal below, access to the heating pipes for maintenance staff and the visual obstruction created by such a bridge for locomotive drivers. Upon completion in 2011, with the addition of a central walkway for internal access and brushed metal cladding to avoid glare from the sun disrupting railway staff, the bridge marked the start of saving a significant quantity of carbon dioxide from being released by producing heat and power separately.

With a laser-cut, modern finish, the tube-like bridge arguably blends in with the many public art features within the university campus, but more importantly demonstrates the establishment's environmental conscience. The bridge also stands as one of the few created purely for carrying heat, in the form of steam, adding a much more contemporary purpose to the repertoire of Britain's bridges.

Chapter 39

Scale Lane Swing Bridge, Kingston upon Hull

Kingston upon Hull's association with bridging often defaults to that of the Humber Bridge, some 5 miles from the city. However, with over fifteen bridges spanning the waterway for which it is named, it is more intimately associated with bridging than most assume.

The River Hull has acted more like a canal since being declared navigable in the thirteenth century and, as it gets closer to meeting the Humber Estuary at Kingston upon Hull, has been industrialised considerably over time. This was largely down to the

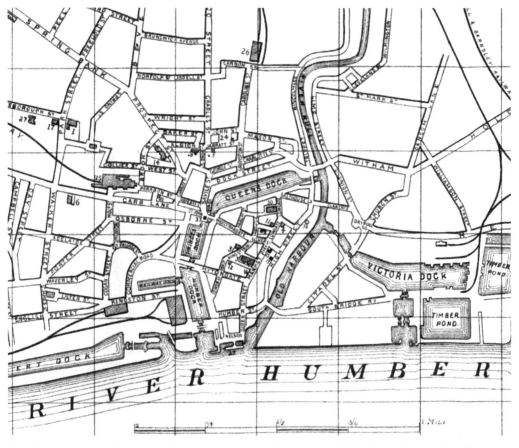

On this map from 1889, Scale Lane's bridge would be in the Old Harbour. (Author's Collection)

importance of transport in and around the North East, which enabled the growth of mills, quarries, farming and, later, coal mining and the fishing industry. To ensure that the Port of Hull and surrounding town could still operate around the river, a number of moving swing and bascule bridges were installed.

By the twenty-first century, much of this need for waterborne transport has dwindled, but the need for the city to overcome the obstacle of a river, in and among the growing levels of pedestrian traffic, warranted the construction of another bridge where Scale Lane met the riverside. The main concern was to link the historic Museum Quarter with an area hoping to spark redevelopment around the busy port, but to satisfy the legacy of hundreds of years of water traffic the bridge had to be movable. The result was the Scale Lane Swing Bridge, a striking architectural feat opened in 2013 that allows the flow of traffic on the river to continue with that of pedestrians and cyclists.

In a rather novel fashion, the bridge allows those on foot or bicycle to stay on the bridge as it opens, reportedly a unique feature worldwide and one that caters well to the desire to bring tourists to the city. Shortly after completion, Hull was announced as the 2017 UK City of Culture, and a number of interesting landmarks and artistic placements matched the contemporary design of the River Hull's latest span. Despite its unusual shape, the design was born from the practicalities of bridging a river that is only 50 per cent navigable (the other half being mud banks). This resulted

A sequence of images showing the movement of the swing bridge. (Timothy Soar)

in the 'comma' shape, with a large structure built into the mud on the west side, counterbalancing the cantilevered point that allows small craft to travel under it, as well as being easy to pivot when larger craft require passage. Architects McDowell and Benedetti, along with engineers Alan Baxter Associates and Qualter Hall, ensured that this form was maximised by allowing pedestrians to enter and exit the larger end while the bridge was in motion, as well as bringing in revenue via a rotating restaurant and viewing deck.

The people making use of the bridge lie at the heart of the bridge's purpose outside of its use for water traffic. A number of artistic projects have been hosted along the span since its opening, such as allowing the public to 'play the bridge' as a giant musical instrument and hosting the 'Sea of Hull', which comprised a thousand nude people gathering along its length. As a result, the bridge has become a centrepiece of creative expression in line with the city's ambition to shed its tired, grey persona. A soundscape installation along the bridge continues this goal permanently. How Hull's future will develop following its year in the spotlight is arguably uncertain, but the bridge at Scale Lane will no doubt exist for decades to come as a modern landmark in the eclectic city.

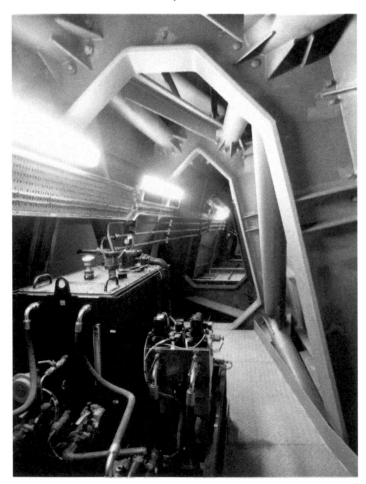

The 'spine' of the bridge gives maintenance staff access to its inner workings. (Timothy Soar)

Chapter 40

Queensferry Crossing

The demise of the Forth Road Bridge began during the dawn of the twenty-first century when it became apparent that the structure's traffic limit was being exceeded for more than half of each year. With almost double the amount of vehicles for which the bridge was planned passing over the Firth of Forth, serious questions were raised as to whether it would withstand the 120 years of service predicted upon it opening in 1964.

By 2011 construction had started on a replacement, which had been proposed and then considered more seriously in the preceding years and warranted a thorough and detailed analysis of the previous bridge's cables. Initial estimates placed the year for the need to build a new crossing at 2020, but with high-tech acoustic monitoring concluding there had been a 10 per cent loss in cable strength, and with other bridges of similar construction beginning to deteriorate in the United States, it became clear that an alternative was required far sooner than first thought.

Project Architect for the Forth Replacement Crossing, as it was then known, was Christian Ernst, who was already known in Scotland for his involvement with the Tradeston pedestrian bridge in Glasgow city centre. With the project across the Forth, a cable-stayed design was implemented, with three towers that would make a gentle reference to the Forth Rail Bridge

Above: *The three piers stand impressively at night, prior to being connected by sections of roadway. (Alan Frew)*

Right: *At South Queensferry, the opposing lanes of traffic are initially separated by the large V-shaped pillars beneath. (Author)*

less than a mile away. By 2016 these towers had taken shape and looked oddly reminiscent of the three monumental cantilevers of its Victorian rail-carrying partner. Despite being leaner in appearance, and with much less framework, on completion the Forth's latest bridge measured a total length of 1.7 miles and the towers topped at 160 meters, making it not only grander than the rail bridge, but also the UK's tallest bridge.

Much of the road deck and other sections were constructed in China and mainland Europe, highlighting the need for such endeavours to make use of international collaboration, but some elements were home-grown, with steel fabrication coming courtesy of Darlington. Notably, no contracts for construction were awarded to Scottish firms. In addition to contemporary construction methods being used for the structure, much focus lay within the development of a 'smart' highway, including the implementation of intelligent lighting, cloud-based maintenance scheduling and updates to the Traffic Scotland National Control Centre.

On 4 September 2017, exactly fifty-three years after the opening of the Forth Road Bridge, and now with an extensive back catalogue of bridge openings, Queen Elizabeth II formally opened the Queensferry Crossing, commenting on the spectacle that now existed courtesy of three iconic bridges spanning three centuries. Those involved with the bridge's conception were quick to point out the economic benefits of rerouting the A90 from the old bridge, upgrading it in the process to its new guise as the M90 and facilitating heavier traffic flow between Edinburgh and the Highlands. With the Forth Road Bridge now vacant, plans were put in place to continue its use for public transport, cyclists and pedestrians via the newly designated A9000.

As one of Scotland's largest infrastructure projects, the Queensferry Crossing appropriately continues a legacy of great bridging achievements across the country's famous body of water, and provides a perfect example of the notion that Britain's future bridges can be as bold, innovative and spectacular as those from an impressive and far-reaching past.

Above: *Changeable weather in the Firth of Forth sees each crossing cope with shrouds of fog, battering winds and, at times, spectacular sunsets. (Alan Frew)*

Left: *The older Forth Road Bridge now lies almost bare, with little more than local buses and keen cyclists crossing it while work is done to replace major components and further weatherproof existing ones. (Author)*